MONOGRAPHS IN COOL CLIMATE VITICULTURE – 1
PRUNING and TRAINING

David Jackson

Daphne Brasell Associates and
Lincoln University Press

Revised edition 2001 by
Daphne Brasell Associates Ltd, PO Box 12 214
Thorndon, Wellington, Aotearoa, New Zealand
www.brasell.co.nz

First published in 1997 by Daphne Brasell Associates Ltd
in association with Lincoln University Press (Soil,
Plant & Ecological Science Division),
PO Box 84, Lincoln University, New Zealand.
Reprinted in 1998 by Daphne Brasell Associates Ltd.

© Text: Soil, Plant & Ecological Science Division,
Lincoln University
© Edition 2001: Daphne Brasell Associates Ltd

No part of this publication may be reproduced, stored in
a retrieval system or transmitted in any form or by any
means, electronic, mechanical, photocopying, recording
or otherwise, without prior written permission of the
author.
ISBN: 0-908896-54-9
Acknowledgment: The author thanks Marion Trought
for many of the diagrams

Cover photograph: PHOTOSOURCE
Production supervision by Jo Fisher, Heather Gamble,
Joan Halliwell, Jane Hingston and Sven Wilson,
Whitireia Publishing, Auckland
Typeset by: Chris Judd, Auckland
Printed by: Astra Print Ltd, Wellington

Contents

	INTRODUCTION	1
1	**IMPORTANT DEFINITIONS**	2
	Cool climate viticulture	2
	Alpha and beta zones	2
	Training systems	3
2	**BASIC PHYSIOLOGY**	4
	Inputs for growth and development	4
	Photosynthesis	4
	Light quality	5
	Light quantity	5
	Water and mineral uptake	5
	Transpiration	5
	Translocation	6
	Respiration	6
	Metabolism	6
	Vegetative growth	6
	Apical dominance	8
	The end point principle	8
	The trunk proximity principle	9
	The highest point principle	9
	The early growth principle	10
	The root-shoot principle	10
	The principle of the node-trunk ratio	10
	Growth patterns	11
	Strategies for survival	11
	Tendrils	11
	Vigorous shoot growth	12
	Vigorous root system	13
	Initiation of inflorescences and flowers	13
	Light intensity	13
	Temperature	13
	Position of buds on the cane	13
	Fruit set	13
	Light and shade	14
	Rainfall and cool temperatures	14

	Fruit development and maturation	14
	Sugar levels	14
	Speed of maturity	15
	pH and potassium	15
	Colour development	15
	Other flavour components	15
3	**CROPPING POTENTIAL OF VINES**	16
4	**LIGHT INTERCEPTION AND UTILISATION**	18
	General	18
	Vineyard configuration or geometry	18
	Canopy height and distance apart of rows	19
	Direction of rows	20
	Other configurations	22
5	**FACTORS INFLUENCING CHOICE OF PRUNING/TRAINING STRATEGIES**	24
6	**SPUR AND CANE PRUNING**	25
	Relevant factors	25
	Spur pruning	26
	Cane pruning	26
	Head cane with canes both sides	27
	Head cane with canes one side	28
	The double head	29
	The cordon cane	29
7	**VINE SPACING**	30
	High-density planting	30
	Low-density planting	30
8	**DETERMINING BUD NUMBERS**	32
	Classical and New World vineyards compared	33
	Magic numbers	33
	Estimating bud numbers	34
9	**TRELLISING**	35
	Trellises in the single vertical plane	37
	The Single Upright 1: Vertically shoot-positioned	37
	The Single Upright 2: Using downward-pointing shoots	42
	The Curtain or High Sylvoz	42

	The Single Upright 3: Using upward- and downward-pointing shoots	44
	Mid-height Sylvoz: Version A	44
	Mid-height Sylvoz: Version B	45
	Mid-height Sylvoz: other versions	45
	The Scott-Henry	46
	Trellises in the double vertical plane	48
	The Geneva Double Curtain	48
	The Lyre	50
	The Ruakura Twin Two Tier	52
	Growing on a pole	53
	Making a decision	54
10	**TROUBLE SHOOTING**	55
	Shoots variable in length and vigour	55
	Uneven shoot distribution	56
	Vines always too vigorous	56
	Trunk girdling	56
	Chemicals	57
	Control by cropping	58
	Root pruning and restriction	58
	Competitive crops	58
	Shoots killed by frost in spring	58
	Persistent suckers and shoots	59
	Badly bowed trunks	60
11	**ESTABLISHING AND TRAINING YOUNG VINES**	61
	Planting material	61
	Ground conditions	61
	Wind, water and weeds	62
	Planting	62
	Training young vines	63
	The trellis	64
	REFERENCES	65
	INDEX	66

Introduction

It has often been said that there are as many training systems for vines as there are people doing that training. It might be also said that books or articles written on pruning and training are as diverse as the authors themselves. Nevertheless, while books necessarily repeat many common aspects, most take a particular view that, in many cases, can be complementary to those published elsewhere. I hope this book is no exception.

I have written it from the standpoint of cool climates in the belief that such climates bring with them specific benefits and problems that can affect the way we approach pruning. You will notice that I have adopted a number of new approaches to analysis and interpretation of pruning and training in the hope that these will add to our understanding of the vine and its management.

To some extent this is a detailed elaboration of the pruning and training section of an earlier book (Jackson & Schuster, 1994).

David Jackson
October 2001

1 Important Definitions

Before getting too far into this monograph it is necessary to define several important terms.

Cool climate viticulture

In areas we normally classify as cool, grapes tend to ripen just before winter sets in. At this stage, leaves are beginning to assume autumn tints, and frosts, which might lead to premature leaf drop, are a worry. For most such areas a warm season is a bonus because ripening can proceed without a hitch and wines are of good quality. A cold season, by contrast, is unwelcome because the grapes may not ripen adequately and wines tend to be of lower quality. Growers in cool climates will choose warmer mesoclimates* within the district and adopt strategies to advance maturity. In warmer climates, grape production is not so constrained because adequate ripening will still occur in cooler seasons, a situation that may, in some circumstances, actually be desirable.

This difference is probably the crucial factor distinguishing cool from warm climates. In the former, lower temperatures before harvest slow down the attainment of maturity and probably create the characteristics that distinguish cool climate wines – lower alcohol, lighter body, fruity aromas and more dominant acidity. They also determine the management practices that growers will adopt to optimise the chances of making good wine. For this reason, Jackson and Lombard (1993) suggest that a useful way to distinguish warm and cool climates is to consider the temperature in the final month before grapes are normally picked in those areas generally considered cool – for example, French areas north of Bordeaux, Germany, Western American districts north of California, parts of Southern Australia, and most of New Zealand. In those areas the temperature in the month before harvest is almost always below 15°C. We now use this to define cool climates

Alpha and beta zones

Some districts have characteristics of both warm and cool climates. In Hawkes Bay, on the eastern coast of the North Island of New Zealand, Müller Thurgau ripens towards the end of March, and the temperature in the month preceding maturity is 16.8°C, by our definition a warm climate. Cabernet Sauvignon ripens at the end of April. The mean temperature for that month is 14.2°C — a cool climate. Thus, according to the variety grown, the climate is either cool or warm.

To avoid confusion, we have coined the terms *Alpha* and *Beta* Zones, which are defined as follows:

> An area is an alpha zone for a specific variety if the temperature in the month preceding harvest is below 15°C.
>
> An area is a beta zone if the temperature is 15°C and above.

Continuing the Hawkes Bay example, we can expect that Müller Thurgau will ripen adequately in most, if not all, years and that growers will not need to adopt techniques

---

* **Macroclimate** – the general climate found in an area; **Mesoclimate** – a modified part of the macroclimate. Mesoclimates may be formed by geographical features, adjacent trees, and/or soil management techniques; **Microclimate** – the climate within or near the vine. It may be induced by pruning and training practices.

(described later) to advance maturity. We could expect Cabernet Sauvignon (beta zone) to have more vintage variation and the effects on wine quality of cool and warm seasons to be more pronounced. Growers can usefully apply techniques to advance maturity under these conditions.

The value of the terms alpha and beta zones is that they give growers specific information on the likely behaviour of individual grape varieties in their districts. However, the terms cool and warm climates can still be used in a more general way. We would define a cool climate as one that is an alpha zone for the major varieties grown in that district. A warm climate is one that is beta for the major grapes in cultivation. Cool climates so defined equate to areas with degree days below 1350 (Gladstones, 1992) or latitude temperature index (LTI) below 380 (Jackson & Cherry, 1988).

One of the most critical factors we need to recognise when pruning and training vines in cool climates is that of lateness in maturity. Pruning and training can affect maturity and other vine conditions in several ways:
- Too many buds left at pruning time may increase yield and delay maturity.
- Training that encourages congestion in the canopy and shading of leaves and fruit can have deleterious effects on quality in both cool and warm climates. In cool climates, shade means the average temperature of the bunch is lower and ripening will be delayed, a consequence of less significance in warmer areas.
- Training systems that do not give good light penetration can have the serious effects of inducing poor flower initiation and poor fruit set. Initiation and set are better in warmer than in cooler areas, making the consequences of poor set or initiation greater in cool climates.

We will investigate further these and other aspects of the interplay between climate and pruning throughout this book.

Training systems

A *training* system refers to the way we position the vine in space. It is an all-embracing term and incorporates (i) the *trellis* on which the vine is grown and (ii) the way we manipulate or train the vine to cover the trellis.

Training in itself can be subdivided into *pruning* and *positioning*. Pruning is the cutting of shoots and canes and may refer to the pruning of dormant canes, usually with secateurs (called 'pruners' in the United States), and normally with some precision. Mechanical winter pruning is a less common and less precise technique. Pruning done on summer shoots, most commonly with machines, is known as 'trimming', which is the cutting of the sides as might be done with a hedge, or 'topping', where just the tops of shoots are cut. Positioning refers to the operations we use to ensure that the growing shoots are correctly spaced in the canopy. It usually involves a moderate amount of handwork and is not easy to mechanise.

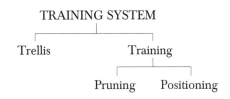

The title of this monograph, 'Pruning and Training', is chosen for its familiarity rather than its precision.

2 Basic Physiology

This section includes a brief summary of basic plant physiological principles together with some more detailed physiology of particular relevance to vines and their pruning and training.

Inputs for growth and development

Grapes need four major inputs for growth and development – water, minerals, carbon dioxide and oxygen.

- Water is vital for all life. Almost all water for plants comes from the roots, although some can be absorbed by leaves and other organs.
- Minerals include the major elements: nitrogen, potassium, phosphorus, calcium and sulphur. The minor elements include iron, magnesium, manganese, zinc and boron.
- Carbon dioxide is absorbed by plants through the leaves. Once there it may combine with water to produce carbohydrates under the action of light. This process is called photosynthesis.
- Oxygen is released during photosynthesis – see equation following. This input may be used in the plant or released to the surrounding atmosphere.

Various plant processes utilise and coordinate these inputs. They are photosynthesis, water and mineral uptake, respiration and metabolism.

PHOTOSYNTHESIS

Photosynthesis occurs when sunlight falls on plant leaves and other green tissue. Green cells, called chloroplasts, are activated by light (the energy source), which leads to the conversion of carbon dioxide and water into sugar and oxygen. The following equation describes the overall process of photosynthesis:

$$6CO_2 + 6H_2O \xrightarrow{photosynthesis} C_6H_{12}O_6 + 6O_2$$

This means that carbon dioxide and water combine, under the action of photosynthesis, to produce simple sugars (glucose and fructose) and oxygen.

Carbohydrates contain simple sugars such as fructose, glucose and sucrose and more complex compounds like starch, glycogen and cellulose. Simple carbohydrates are used by the plant for energy (see later) and as building blocks for the production of more complex carbohydrates plus many other materials needed to construct the plant body. Likewise, they are used by other organisms – animals for example – which, directly or indirectly, consume and utilise plant parts to supply their own energy and construct their own bodies.

We do not intend to discuss the complex biochemistry of photosynthesis here, but the ramifications of the process are important to understand.

FIGURE 1: Photosynthetic value of light at different wavelengths.

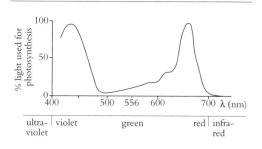

Light quality

White light is composed of coloured light ranging from far-red to ultraviolet. Figure 1 shows the percentage of the light at different wavelengths that is absorbed by the leaves and used for photosynthesis. Note that very little light in the centre of the spectrum (green) is absorbed by the leaf. (Leaves appear green because such light is not absorbed but reflected.)

The viticulturist normally has no control over the nature of the light received by the vineyard. As light passes through the canopy of leaves its composition changes. Light of longer wavelengths becomes more dominant. This occurs because leaves absorb more of the short wavelength light. Evidence from various sources (Archer & Strauss, 1989; Smart, 1987; Smart, Smith & Winchester, 1988; Morrison & Noble, 1990) has shown that long-wavelength light (red, far-red) can modify the composition of grape juice and produces grapes with high pH and potassium levels.

Light quantity

Light intensity is usually measured as electro-magnetic energy, namely $\mu Mm^{-2}s^{-1}$, which is the same as $\mu Em^{-2}s^{-1}$ (μM = micromol, μE = micro Einstein). Sunlight can be as high as 2000–2500 μM, but in cloud and shade it may be much lower. A dull, overcast day may be in the order of 200–300 μM. The vine cannot utilise more than 700–800 μM. At 28 μM the amount of carbohydrate produced by photosynthesis equals approximately that consumed by respiration (discussed shortly). Below this intensity, leaves turn yellow (senesce) (Figure 2) and may fall off (abscise). Low light intensity produces the following effects (Jackson & Lombard, 1993):
- *on leaves*: grape juice (usually called must) tends to have high pH and acids, and low sugars;
- *on berries*: tends to delay maturity and impairs colour development.

FIGURE 2: Shaded leaves in centre of vine.

These consequences tend to reduce the quality of wine made from affected berries. This reason is why we emphasise in this monograph the importance of canopy management to reduce shading.

WATER AND MINERAL UPTAKE
Transpiration

Water is taken up from the soil predominantly by a process called transpiration. In essence, transpiration is a mechanism whereby water, which is lost from the leaves, is replaced by water drawn into the roots. The water travels upwards, partly by capillary action, in the longitudinal vessels in the wood (xylem). The leaf, usually on the undersurface, has specialised structures called stomates (or stomata). These are pores through which water vapour can escape, creating the negative pressure that drives transpiration. In situations of high stress levels where dehydration is possible, stomata are closed by the modification in shape of two cells (guard cells) that surround the pores.

Transpiration moves water and dissolved minerals throughout the plant, and loss of water by evaporation acts to cool the leaves and the plant under hot conditions. Air exchange through the stomata enables the process of photosynthesis to occur. If water stress occurs, closure of stomata effectively stops photosynthesis and prevents the cooling effect of loss of water vapour.

Translocation

Translocation is a more specific mechanism used to move minerals in the plant. As already seen, transpiration moves solutes in the plant (solutes are minerals dissolved in water), but this process does not discriminate between different minerals. The minerals in the soil, or more correctly the solutes surrounding the roots of the plant, are not in the same concentrations as are required in the plant tissues. Translocation acts to redress this balance. Just as transpiration provides a negative pressure to suck up the liquid, translocation creates a positive pressure from the roots into the plant as a whole. Evidence of this positive pressure is seen in early spring when vines are pruned before bud burst; this forces the sap upwards, causing bleeding through cut branches. The cells of the root are able to discriminate selectively between different elements and provide a solute mix appropriate for the plant. Translocation is an energy-driven process.

RESPIRATION

The process of respiration is essentially the reverse of photosynthesis, namely:

$$C_6H_{12}O_6 + 6O_2 \longrightarrow 6CO_2 + 6H_2O + E$$

Simple sugars (glucose and fructose) with oxygen release carbon dioxide, water and energy.

Energy can be used for the diverse processes needed by the plant, for example, translocation, production of new compounds for use in developing the plant's structure, and the laying down of storage materials for later use.

Respiration requires the presence of oxygen. Anaerobic conditions (such as roots in inadequately aerated soils) cause other undesirable products to be formed that may result in plant death.

METABOLISM

This term refers to the changes in chemical composition within plants (and all living organisms). These changes include the building up of the constituents of the organism and the breakdown of molecules to provide energy. It is an all-embracing term that includes photosynthesis and respiration. The products of metabolism are sometimes called metabolites.

Vegetative growth

Shoots of grapes grow from spring to autumn by the proliferation of nodes* and internodes at the apex (tip). Each node or swelling can produce one leaf, one bud and one shoot. The bud is, in fact, compound and has three growing points that produce the shoots and inflorescences the following spring. Normally only one growing point – the primary bud – develops into a shoot, but occasionally two or, rarely, three develop. If a primary shoot is damaged, the secondary bud may grow. Likewise, if the secondary is destroyed, the tertiary one will grow.

The third structure at the node is called a lateral shoot. Laterals may make minimal or extensive growth according to the vigour of the plant and according to whether the main shoot is allowed to grow unhindered or whether it is topped or damaged, in which case lateral

---

***Node** – the swelling at the point where the leaf stalk (petiole) meets the stem; **Internode** – the portion of the shoot between two nodes (does not have the capacity to produce buds, leaves or shoots); **Inflorescence** – the cluster of flowers which, after flowering, is called the bunch.

FIGURE 3: *Top*: young shoot four to five weeks after bud burst. *Bottom*: young lateral shoot and dormant bud in axil of leaf. The bud is a compound bud with three growing points – primary, secondary and tertiary.

FIGURE 4: *Top*: a vigorously growing shoot has tendrils that extend past the apex. *Bottom*: a shoot where growth is terminating. The tendrils do not extend past the apex.

growth is encouraged. Buds burst in spring, and the developing shoot has the appearance indicated in Figure 3.

After the third or fourth leaf has appeared, an inflorescence forms at the next node on the opposite side to the leaf. Two and sometimes three inflorescences may be produced on consecutive nodes. The next node has no appendage (apart from a leaf, bud and possibly a shoot), but the one after that has a tendril where previously inflorescences were situated. Tendrils then form on alternate nodes in *Vitis vinifera* but slightly differently in some other

Basic Physiology

grape species. Shoots grow until some factor prevents their further development, at which stage the shoot tip drops off (abscises). The shoot is then terminated by a leaf and the bud in the axil of that leaf.

A vigorously growing shoot has long tendrils that extend past the tip of the shoot. A less vigorous one, or a shoot where growth is about to stop, has weak tendrils that do not extend past the tip (Figure 4).

APICAL DOMINANCE

A characteristic of growing shoots is that the apex dominates those shoots that originate at lateral positions below the apex. Thus, while the grape shoot grows from the apex, the laterals, described above, may show limited development. If they do develop, they tend to be longest at nodes not too close to the apex. If the apex in an actively growing shoot is severed, apical dominance is broken and vigorous growth of laterals can occur. Later, one of these laterals may assume dominance. If the apex ceases to grow because of such factors as drought or nutrient deficiency, lateral growth is not stimulated.

Several principles govern the vigour of individual shoots, and are partly a response to apical dominance. These principles are described in the following six points.

THE END POINT PRINCIPLE (EPP)

The EPP states that shoots at the end of a cane may have a vigour advantage (see Figure 5). This situation is not quite the same as the apical dominance factor described above. There we

FIGURE 5: The End Point Principle (EPP).

FIGURE 6: The Trunk Proximity Principle (TPP).

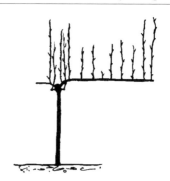
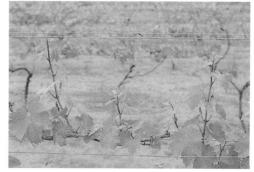

were describing apical dominance on a growing shoot. In this case, the cane was formed the previous season and the shoots have arisen from buds along the cane. The end buds have a growth advantage over those closer to the base.

THE TRUNK PROXIMITY PRINCIPLE (TPP)

This principle states that shoots that originate close to the trunk may have a competitive advantage over other buds and will usually grow more vigorously (Figure 6). This situation seems to be the opposite to the EPP. In fact, the above situation may not occur in a cane-pruned vine due to the fact that the EPP is stronger than the TPP. Growers often notice vigorous shoots near the trunk, but these are usually seen to originate from replacement spurs. In situations where too many buds on a cane have been left, the TPP is more noticeable (see also under the root-shoot principle below).

Evidence of the TPP can often be seen when *spur* pruning is adopted (we will discuss cordon and cane pruning later) although, as a general rule, shoots from spur-pruned vines tend to have more even growth than cane-pruned vines.

THE HIGHEST POINT PRINCIPLE (HPP)

The HPP indicates that the shoot at the highest point can also have a growth advantage (Figure 7). Sometimes a cane arched in the centre will modify the EPP to achieve more even growth.

Gravity is probably involved in the HPP. Gravity causes a redistribution of plant hormones in the cane between the top side and the underneath side in such a way as to favour the former. Thus, with individual buds on canes, the one on the upper side will be more likely to grow than one on the under side.

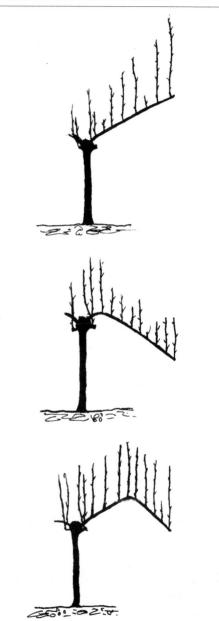

FIGURE 7: *Top and middle*: two examples of the Highest Point Principle. *Bottom*: an arched cane used to even out the EPP and TPP.

Basic Physiology

FIGURE 8: The root-shoot principle. *Above*: heavily pruned vine – shoots tend to be more even in length. *Below*: lightly pruned vine (TPP left; EPP right).

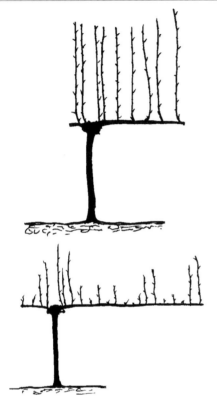

THE EARLY GROWTH PRINCIPLE (EGP)

This principle states that a shoot which begins growth early will tend to keep its competitive advantage.

Note that we have been a little less than dogmatic in describing these principles, using terms like 'tend to', 'may', 'usually'. We have done this because other factors may influence application of these principles in a specific instance. One particular factor is the root-shoot principle.

THE ROOT-SHOOT PRINCIPLE (RSP)

This principle states that the ratio of shoot mass to root mass tends to stay the same. For example, after a season's growth, the root/shoot ratio, especially in a vine not summer pruned, will be close to that which the plant would normally try to achieve. If the grower prunes the plant hard, the plant tries to restore that ratio. The shoots tend to grow vigorously, with each shoot usually being of similar vigour and the responses described above being minimised (Figure 8).

However, if a lot of buds are left on the vine, especially a weak one, the effects may be accentuated. In such circumstances the TPP occurs commonly on cane-pruned vines.

THE PRINCIPLE OF THE NODE-TRUNK RATIO (PONT)

While the plant is young, the above principles will be exaggerated (an undesirable situation) if the ratio of the node numbers (left after pruning) to trunk diameter is too high. As a guide, if the trunk diameter is from 5–15 mm, the number of nodes (buds) that can be left is approximately 1.5× the diameter of the trunk in millimetres. Thus, if the diameter is 6 mm, the bud number should be 9; if 12 mm, then 18. Once the diameter reaches 15–20 mm, then, providing planting distance is not too large, the plant will be filling the canopy and the crop will be close to maximum, which means that the principle of the node/trunk ratio will not operate with any precision.

The names used for these principles are my own and will not be found in earlier books on pruning. Their value is in identifying and discussing vegetative growth of vines.

GROWTH PATTERNS

Grape varieties vary in many characteristics such as leaf shape, bunch shape, yield potential, etc. One aspect that has special importance for pruning and training is the nature of shoot growth. Some varieties tend to be very upright, others tend to be floppy. This can be a factor that might influence our choice of training system – upright canes are easier to train on, for example, the vertical shoot-positioned canopy (VSP – see later). Floppy canes adapt well to hanging curtains such as the Geneva Double Curtain.

FIGURE 9: Wild *Vitis californica* in North America. The vine must compete with other plants to gain a place in the sun.

Strategies for survival

Every species, whether plant or animal, has evolved strategies for survival. These strategies enable the individual to survive in competition with many other species while coping with the normal environmental features found in the district of its origin. For example, the thorns of roses may reduce damage by browsing animals, allowing them to gain an advantage over their neighbours. The ability of rabbits to run fast and breed rapidly means they can outrun predators most of the time and reproduce rapidly to replace those that are caught. Grapes, likewise, have strategies.

TENDRILS

While lacking a supporting trunk, such as utilised by an apple, pear or an oak to allow them to move above the ground for a place in the sun, grapevines have evolved an alternative method of doing the same thing; they produce tendrils.

Tendrils enable the plant to cling to upright structures and gain height. As such, their young shoots are raised above browsing animals, and their leaves are exposed to the sun at the expense of smaller plants closer to the ground (Figure 9). Such a strategy is common for all vines, although some details may vary. The kiwifruit, for example, has no tendrils, but young shoots in the proximity of a potential supporting structure will wrap themselves around that structure to gain support.

The clear advantage of tendrils or twining shoots to vines when growing in the wild are not necessarily beneficial to vines under cultivation by humans. In some places, they are grown in association with trees, which provide support, but mostly they are produced under monoculture. Humans provide the structure (a trellis) and proceed commonly to guide and support the shoots in the position they are to

Basic Physiology 11

be held. In such circumstances, tendrils are redundant and even a nuisance because they increase the difficulty of removing old canes from the trellis after winter pruning.

VIGOROUS SHOOT GROWTH

As woody plants grow from seedlings or cuttings to maturity, they normally moderate their growth rate. Young apple trees, for example, produce vigorous shoots in their early years, but as the trees mature, shoot growth declines. In fact, the shoot normally terminates its growth in late spring or early summer while shoots of young plants may grow for most of the season. This growth pattern makes survival sense. The tree needs to grow rapidly to establish a place in the sun. Once there, it must only maintain its position and reduced extension will suffice, which means that less energy goes to the shoots and more to the trees' reproductive organs, that is, their flowers and fruit. Vines are different in that the mature plant maintains its capacity for vigorous growth: sometimes 5 m of vine growth may occur in contrast to 50–100 cm in, say, a mature apple tree.

Once again, the advantage to the plant in the wild can be a disadvantage to growers. In natural conditions, the vine must make more growth than its host so that it can quickly reach the sunlight from the shaded understorey. Once there, its growth must still be more vigorous than the host, so that it can maintain its position.

Early rapid growth *does* have advantages – the vine quickly fills its allotted space and vineyards normally come to full production earlier than most fruit orchards. But from this point onwards excessive vigour is a problem. Ninety percent of the growth that the plant made during the previous season is commonly removed by the pruner in winter. Even summer trimming of the vine may leave less than 30

FIGURE 10: The number of inflorescences will already be formed in the dormant bud (top). The flowers (below) form on the inflorescences in spring.

percent of the potential shoot volume that could be produced. In other words, the plant is using many times the energy it needs to achieve an adequate leaf area. This process is clearly inefficient and is probably one reason why grapes characteristically produce less per hectare than other fruit crops (a grape yield of 20 t/ha would be matched by an apple crop of 80 t/ha).

VIGOROUS ROOT SYSTEM

The third survival strategy is the vigorous and deep root system common in vines. Once established, such a root system enables the vine to survive conditions of drought stress that many other plants could not tolerate. For grape growers, this can be an advantage in many sites, but if the soil is deep and fertile and adequately supplied with water, it can contribute to the vigour problem discussed above.

Initiation of inflorescences and flowers

As noted earlier, buds that burst in spring form during the previous growing season. In addition, within these buds, inflorescences (primordial bunches) are present by the end of the growing season. Dissection of a bud at this time will reveal three to four primordial leaves followed by, usually, two inflorescences opposite the next two leaves. In spring, buds swell and flowers form on the inflorescences. When they emerge, the flowers are already extant (see Figures 3 and 10).

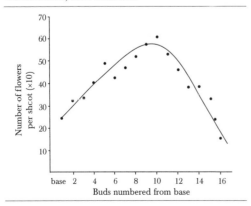

FIGURE 11: Fruitfulness of buds from the base of a cane: mean figures for 10 cultivars in Christchurch, New Zealand.

Three major factors influence the ability to produce a good complement of inflorescences and flowers, often called 'fruitfulness'. These are light intensity, temperature and the position of the bud on the cane.

LIGHT INTENSITY

There is a relationship between light intensity and the number of inflorescences and flowers. Buds will produce more flowers if, in the previous season while they were being formed, more light was received by the bud and its adjacent leaves. The method used to prune and manage the shoots that grow is one of the major factors growers can use to improve flower formation.

TEMPERATURE

High temperatures in the early part of bud formation increase the number of inflorescences that form. We have a reduced ability to modify this situation once a site has been chosen. However, shaded shoots will be cooler than exposed ones, so factors that improve light penetration to the buds will have the secondary effect of increasing temperature and inflorescence number.

POSITION OF BUDS ON THE CANE

Buds closer to the base of the shoot and at the tip are less fruitful than those in the centre (Figure 11). This factor is important in determining whether growers cane prune or spur prune (discussed below). Generally, in cool climates, where temperatures are low during the period when inflorescences are formed, buds are less fruitful. It is therefore more important that growers retain a high proportion of the central buds, as occurs with cane pruning.

Fruit set (coulure)

Once the inflorescences and flowers have formed, the next obstacle to producing an

adequate crop of berries occurs during the conversion of flowers to fruits. This conversion process is known as fruit set (also called coulure), and it is affected by:
- *early bunch-stem necrosis (EBSN)*: This refers to the shrivelling and death of sections of the bunch occurring between the time the bunch is 2 cm in length and anthesis (the shedding of pollen by anthers),
- *the ability of flowers to be effectively pollinated and fertilised*: The important time to influence pollination and fertilisation is during the period of capfall (petal fall).

A low incidence of EBSN and good pollination and fertilisation means that growers can celebrate a 'good fruit set'. Conversely, 'poor fruit set' may be due to high EBSN or poor pollination and fertilisation, or both. Several factors influence fruit set. The most important, in the context of pruning and training, are *light intensity*, *rain* and *temperature*.

LIGHT AND SHADE

The development of EBSN is considerably exacerbated by heavy shading. In vines where very low light intensity occurs in the inflorescence zone, up to 100 percent of bunches can be affected and massive yield loss may occur. Clearly, good trellis design and canopy management is vital for vineyard operators to reduce such shading, normally caused by too many shoots in a restricted zone.

RAINFALL AND COOL TEMPERATURES

Under cool, wet conditions, fruit set is commonly reduced. Unfortunately, the scientific literature on fruit set does not reliably indicate whether poor fruit set under these conditions is due to EBSN (possibly induced by the attendant low light intensity) or poor pollination and fertilisation. Nevertheless, growers who keep an open canopy to aid rapid drying after rain will minimise the risk.

Fruit development and maturation

During the 1980s, scientists devoted considerable effort to discovering those factors that influence fruit development and maturation, modify juice and wine composition and, ultimately, wine quality. We will look at the production of quality in juice and wine later in this series of monographs, but it is necessary here to anticipate a little of this discussion by mentioning a few key features that can be influenced by growers during pruning and training operations. (Sources for this information can be found in Jackson & Lombard, 1993.)

SUGAR LEVELS

Higher sugar levels will be found in grapes if the leaf/fruit ratio is high. A ratio of 10 cm^2 of leaves per gram of fruit is adequate and will usually achieve sufficient sugar. Above 10 cm^2 little additional increase will occur. Growers should generally aim to achieve this leaf/fruit ratio by ensuring that leaves are fully operative for photosynthesis through good exposure to sunlight.

Later, when examining various training systems, we shall consider the potential leaf/fruit ratios of these systems. Heavily shaded leaves in fact consume more than they produce and will not contribute, and may even retard, sugar accumulation in berries. Sugar accumulation is not automatically linked to quality but, especially in cooler climates, quality wines are more commonly produced from grapes with higher sugars.

SPEED OF MATURITY

So many factors change as maturity approaches that keeping tabs on even a small number is not easy. For simplicity we normally measure rise of sugars, increase in pH, fall in acids and changes in colour. The key factor linked to such changes is temperature. With medium-high temperatures, the rate of change is more rapid. In many cool climates a short season may limit adequate ripening, so growers are particularly keen to maximise temperatures.

Temperatures, as we already know, change with macro- and mesoclimates, but the microclimate within the vines also plays a role. Pruning and training will modify this situation. Fortunately, the practices we have already discussed as being beneficial for flower initiation, fruit set and sugar accumulation provide an optimum environment for rapid maturity. Sun-exposed berries will be warmer than shaded ones and, provided leaves are similarly exposed for adequate photosynthesis, ripening will be facilitated.

pH AND POTASSIUM

Changes in pH and potassium accumulation are influenced by shading. High shading of leaves, especially those close to the bunches, encourages potassium accumulation and high pH. These can reach levels that are inappropriate for quality wine production. Fortunately this is less common in cool than in warm climates.

COLOUR DEVELOPMENT

Grape berries *will* develop limited colour in shade, but the intensity is heightened if exposed to sunlight. As a consequence, as for so many other factors, competent canopy management is essential if good colour development is to be achieved. We might also recall that very high and very low temperatures reduce colour development.

OTHER FLAVOUR COMPONENTS

We still do not know the full details of all the chemicals that contribute to the flavour of grape juice and wine, but even here we have evidence that light exposure makes a contribution. Methoxypyrazines, for example, are the compounds that are significant in imparting a green, 'grassy' aroma to wines. These can become unpleasantly high in many grapes if berries are heavily shaded.

There is some evidence that berries that receive too much direct exposure to high-intensity light may suffer from a too-high level of phenolics.

Basic Physiology

3 Cropping Potential of Vines

The potential yield of a vineyard is probably up to 100 tonnes per hectare if all conditions are optimum (Lavee & Haskal, 1980).

Such optimum conditions are:
- interception of almost all the sunlight falling on the vineyard by leaves and shoots that do not overlap and are not crowded,
- the use of heavy-cropping varieties; low-cropping vines do not have not the genetic potential to produce such high yields,
- the provision of adequate water and nutrients plus appropriate disease control and other management practices, and
- the availability of a climate with high-intensity light, adequate heat and low rainfall (compensated by irrigation).

The reason why wine-grapes do not normally reach the above figure is because:
- not all light is intercepted,
- many wine-grape varieties have low to medium cropping potential,
- optimum soil and water conditions are not selected for cool-climate wine varieties, and
- climates tend to be chosen that are optimum for quality and not for yield.

The most important of these constraints, in relation to the pruning and training regime, is *light interception*, which we have already mentioned and will shortly consider in some detail.

The *varieties* chosen for cool climates tend to crop with only moderate yields in cool climate conditions. Those that do crop more heavily, for example, Müller Thurgau and Breidecker, are usually considered to be moderate- but not high-quality grapes. However, even here, their quality will decline if overcropped.

FIGURE 12: Heavy cropping of table grapes grown in a greenhouse. The high yield is because of (i) ideal temperature conditions, (ii) high light interception, (iii) a very productive variety and (iv) careful crop and leaf management.

A plentiful supply of *water* and a deep, fertile *soil* can ensure the vine will crop to its fullest potential. However, these conditions pose some problems that will normally limit the attainment of this potential. The most serious is the excessive vegetative growth that usually follows from such plenitude. Very careful management and summer trimming are needed to avoid over-shading and the consequent poor fruit set (low yield) and poor quality. Viticulturists often prefer soil and water conditions that limit yield by imposing some stress.

Warmer temperate *climates*, for reasons already mentioned, usually produce heavier cropping. Nevertheless, growers often choose cool climates because of their appropriateness for certain desired wine styles. Such growers need to be satisfied with lower yields but, of necessity, they will need to obtain higher prices per tonne of grapes.

Greenhouse conditions used to encourage heavy cropping can be appreciated from Figure 12.

4 Light Interception and Utilisation

General

In cool climates, the most appropriate way growers can improve yield is to increase the interception of high-quality (that is, not excessively shaded) light. This practice must be modified by two parameters. First, and as noted earlier, for each gram of fruit produced the vine needs 10 cm^2 of adequately illuminated leaf area. For a moderately fruitful variety, this equates to 14 to 16 leaves for a two-bunch shoot. Second, the leaf area we consider must be *exposed* leaf area. A shoot with 15 leaves of which 10 are severely shaded will neither ripen grapes well nor produce a good crop the next year.

The production of adequately leafed vineyards is a matter of geometry. If the vineyard is planted and orientated so that all light is intercepted all the time, it could theoretically give maximum yield. Indeed, it would not be difficult to achieve total light interception. A vineyard unpruned for a number of years would probably be such, but it would be unmanageable and would probably yield much less than optimum for several reasons. Thus pests and diseases may destroy part or all of the crop, heavy shading may induce early bunch-stem necrosis and a search party may be needed to find the grapes amidst the dense growth at harvest. While some training systems do achieve almost total light interception, associated problems usually limit their use. These will become apparent later.

Grapes are grown in three geometric forms:

1) They can be grown separately and vertically on poles, or separately but sprawling as in the Moselle River in Germany or the Goblet system of southern France.
2) They can be grown in rows, either vertically shoot-positioned or sprawling.
3) They can be grown horizontally, such as in a pergola, where overhead foliage is tended from below by the grower.

When the sun is directly overhead, light interception is greatest in the pergola and least in the trimmed separate and vertical system (Figure 13). In fact, no matter what the sun position is, a well-managed pergola will effectively intercept all light and can thus, potentially, give optimum yield.

That said, growers seldom use pergolas, especially in cool climates. (We will consider why this is so later.) The most common system is that of growing in rows with shoots trained vertically and trimmed. This system, termed the Vertically Shoot-Positioned Canopy or VSP, is widely accepted and gives generally satisfactory production of quality grapes. For these reasons, it warrants further description.

Vineyard configuration or geometry

Figure 14 shows the standard VSP trellis configuration in cross-section. When the rows are set north-south, all sunlight is intercepted in the morning and afternoon. At midday, however, the only light to be intercepted is that falling directly on the top; most hits the vineyard floor. When the rows are closer together, interception is increased (Smart, 1973). However, close rows bring other problems, as discussed in the following points.

FIGURE 13: Bird's-eye view of canopy cover resulting from various systems of vine training.

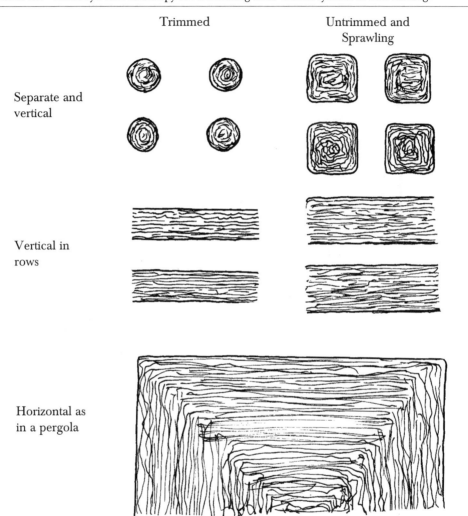

CANOPY HEIGHT AND DISTANCE APART OF ROWS

Although close spacing between rows can potentially increase yield, the light quality in very close plantings may become inadequate, especially towards the base of the foliage canopy. In some of the best vineyards in Bordeaux, rows are 1 m apart and the foliage is 90 cm in height. This configuration intercepts between 40 to 50 percent of the total light available and yet still provides an adequate light environment for yield and quality in that area (Carbonneau, 1979). Generally, the ratio between the height of the foliage canopy and row distance should be below 1.0 and possibly closer to 0.6, especially in cool and less sunny areas.

Light Interception and Utilisation

FIGURE 14: Light interception by a vertically shoot-positioned canopy in the morning, at midday and afternoon.

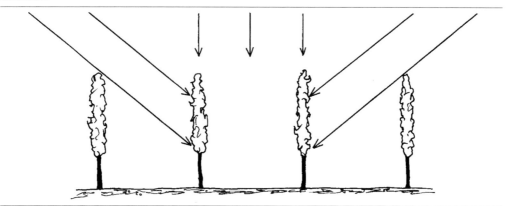

Figure 15 illustrates the ratios for rows 1.0, 2.0 and 3.0 m apart. While theoretically all are equally satisfactory, practical considerations may dictate which is chosen. At 1.0 m apart, special tractors will be required to cultivate and spray. In addition, in a vigorous site, excessive topping may be required to achieve shoots only 0.6 m long. At 3.0 m apart, the structure of wires and posts becomes more complicated and may be expensive, and a divided canopy may be needed to provide 1.8 m of canopy (see later).

DIRECTION OF ROWS

In some properties, the shape and slope of the site determine the direction of rows. Where these characteristics are not a limitation, growers have to decide on the most appropriate direction; generally this is a choice between rows orientated either north/south or east/west

Most vineyards are planted with rows in a north-south direction, and we have already noted the different amounts of light interception at different times of the day.

For east-west rows, morning (and afternoon) sun is the least intercepted; the majority is received toward midday. However, because the sun is high in the sky at noon, not all the light is intercepted, so that for east-west rows there is no time of total light reception.

FIGURE 15: Achieving a ratio of 0.6 (canopy height) to 1.0 (row distance) at three planting densities.

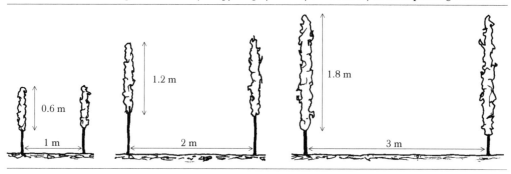

TABLE 1: Percentage of light intercepted by the top and sides of a vertical canopy on a sunny day with rows orientated north-south or east-west at a latitude of 43°.

Light Exposure	21 Dec/June		21 Mar/Sept	
	North-South	East-West	North-South	East-West
a) Light interception at sides of canopy	50	24	52	71
b) Light interception at top of canopy	20	20	20	20
c) Total interception (a & b)	70	44	72	91
d) Light falling on soil surface	30	56	28	9

Source: Adapted from Champagnol (1984) and Smart (1973).
Note: The data is for rows where the height of the canopy is 80 percent of the distance apart of rows. The width at the top of the canopy is taken as 25 percent of canopy height.

Table 1 indicates the percentage of light intercepted in mid-summer (21 December and 21 June, southern and northern hemispheres, respectively) and early autumn/spring (21 March/September) in rows oriented north-south and east-west.

In mid-summer, more light is intercepted by north-south rows, but in spring and autumn, east-west rows receive more light.

Making a choice between north-south and east-west rows is not easy, but one additional factor may predispose growers to north-south rows. This is that both sides of the rows receive approximately equal exposure to light. For east-west rows, one side (south or north depending on hemisphere) receives very little direct light. Generally, therefore, growers choose north-south rows. It may, however, be worthwhile considering occasions where east-west rows might be contemplated. These are as follows:

- *Mid-summer temperatures are very hot*: Under these circumstances less light interception might be advantageous to reduce water and heat stress.
- *The latitude is relatively high*: This factor increases summer interception compared

FIGURE 16: Deviation from the north/south orientation to maximise reception of late afternoon light in the northern and southern hemispheres.

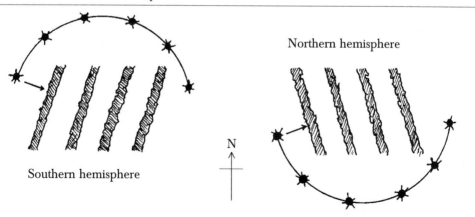

Southern hemisphere

Northern hemisphere

Light Interception and Utilisation

with the low-medium latitude situation shown in Table 1. In addition, the high interception in spring and autumn may enable the plant to get a good early start to the season and achieve better light capture in the ripening days of autumn.

One other aspect of row orientation needs to be mentioned. There is some merit in modifying the direction towards the north-east and south-west in the southern hemisphere and north-west and south-east in the north. This orientation benefits afternoon light capture at the expense of morning capture (Figure 16). Generally, afternoons are warmer than mornings because the soil has accumulated the day's heat and has begun to re-radiate it. Increasing afternoon exposure at the expense of morning exposure means the light has more value to the plant because the warmer temperatures promote more photosynthesis and metabolism and thus assist ripening.

OTHER CONFIGURATIONS

As we have discussed, upright canopy rows allow light interception to increase as the rows are brought closer together. Rows cannot, however, be brought too close together (relative to canopy height) because the lower part of the foliage becomes unacceptably shaded. There are, however, other configurations that are not often used but have clear advantages in light interception. The three we consider here are the overhead pergola, the Tatura trellis and the Lincoln canopy, each of which is depicted in Figure 17.

The overhead pergola potentially uses all light available because coverage is complete. It is high enough to allow tractors and other implements to travel underneath but not too high to prevent workers from pruning and picking from below.

The Tatura trellis was designed to achieve maximum light interception, because the dentate shape and its angle of 60° is considered the theoretical optimum for light utilisation (Van den Ende & Chalmers, 1983).

The Lincoln canopy was developed to achieve the advantages of the pergola, allowing for easier management and a mechanical harvesting potential (Jackson, 1983). Light interception is very high, although not quite as high as with the previous two configurations.

These systems all have one conspicuous problem that prevents their widespread use – foliage management. In the standard upright canopy, the foliage has a specific position in which to grow. In the Lincoln canopy and the pergola, shoots grow upwards and quickly become congested because no provision is made for their positioning. They can be tied down to specific positions, but doing so takes time, or they can be trimmed at a specific height above the bunches, which often results in a poor leaf/fruit ratio.

The Tatura trellis does provide a zone for shoot growth, but training of shoots is more demanding here than with the normal upright trellis. In addition, light interception in the lower part of the trellis tends not to be good. Despite these problems, there are growers who, by good management and careful shoot positioning, produce good-quality grapes from these systems.

FIGURE 17: Three non-vertical canopies.

Pergola

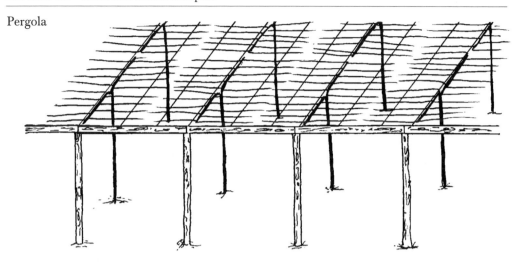

Tatura trellis

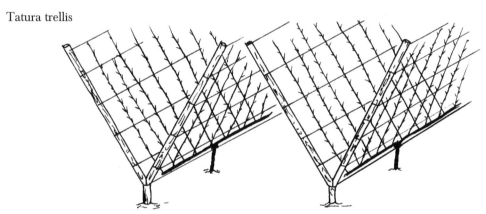

Lincoln canopy

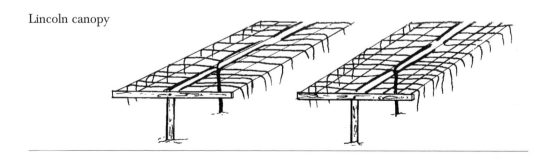

5 Factors Influencing Choice of Pruning/Training Strategies

The most simple and least-complicated strategy for pruning and training is to select the most common local system, and follow carefully instructions for pruning management given by advisers and successful growers in that district.

This approach has the following advantages:
- Its common use probably means it is appropriate and successful in one's own district. There may be better approaches, but growers may not wish to take the gamble of possible failure.
- There are many growers who have the skills and ability to make it work well. Adopting the practice of successful growers is a safe approach, as is the security of having someone's shoulder to cry on if things go wrong.
- Equipment for handling the grapes is probably relatively cheap and widely available. The use of secondhand machinery or machinery-sharing options also can be explored.

The above considerations notwithstanding, in instances where districts have used growing systems that are not the most appropriate, the consequent introduction of new ideas and systems has brought rewards in yield and quality.

6 Spur and Cane Pruning

We have already noted that spur pruning, which retains only the lower buds on a cane, may be less productive than cane pruning. Because grapes in cool climates are often less productive than desired, cane pruning is the more common choice. The following list presents the factors that relate to the two options. (We assume here that the bud number retained per metre of row after pruning is the same in both systems.)

Relevant factors

1) Cane pruning retains buds from the fruitful middle section of the cane, and should increase productivity.
2) It may be easier to determine the buds per metre that will grow as a consequence of cane rather than spur pruning. This is because spurs have at their base one or more buds that *may* or *may not* grow. If we look ahead to Figure 18 (page 26), we notice two separate buds (sometimes called 'count buds'). Below these is another smaller one that is close to the base, and there are other even smaller ones that can, in certain circumstances, form shoots. On a spur like this, the two 'count' buds are expected to grow, but very often the lower one will also grow, and it may be difficult to know whether to estimate two or three shoots per 'two-bud' spur.

   On a cane, the bud normally produces one shoot, although sometimes the secondary and even the tertiary growing point produces extra shoots. It is therefore impossible to be absolutely sure how many shoots will grow, but it is generally possible to predict this outcome more accurately in canes than spurs. Some varieties (for example, Gewürztraminer) are more likely than others to produce multiple shoots from a node.
3) Paradoxically, there is often more chance of gaining the desired bud number per metre with spur than with cane pruning. This seemingly contradictory statement is because spur number per metre and buds per spur can be modified to give a reasonable approximation to the bud number required. If, however, one cane does not happen to give the correct number, it can be doubled or trebled by laying down two or three canes, but a 50 percent increase, say, can only be achieved by adding a short cane, which gives uneven bud density. (This is further discussed on page 56.)
4) A spur-pruned vine retains a higher proportion of older wood after pruning, providing a larger source of stored reserves, a situation that gives more rapid and often more even growth of shoots in spring. Variation through such factors as the End Point Principle (EPP) therefore tends to be less on spur-pruned vines.
5) With age, the spurs may die and gaps may occur on the cordon. New canes may then be required to re-establish the cordon.
6) Blind (non-productive) buds often occur on canes, especially if they developed in shade the previous season. This situation can give gaps in shoot distribution the following spring.
7) Of the two pruning methods, spur pruning is the simplest and easiest to teach to inexperienced workers.

In summary: practice suggests that for adequate cropping in cool climates, cane pruning is normally preferred. Nevertheless in some cool climates with some varieties, cropping can be satisfactory with spur pruning. Therefore growers might begin with cane pruning but experiment with spur pruning on a proportion of the vineyard.

Spur pruning

Spur pruning is the system whereby last year's canes growing from a permanent or semi-permanent cordon are cut back to a limited number of nodes, commonly between two and five. This method leaves a 'spur', the buds on which produce next season's new shoots and crops. (In fact this is not a true spur as one sees in an apple or pear tree, where a shoot makes limited growth and terminates, commonly, with a fruitful bud.) Figure 18 illustrates the method of spur pruning for grapes.

Cane pruning

The basic cuts needed to cane prune a vine are shown in Figure 19. In this system, canes originate from permanent sections of the vine,

FIGURE 18: Spur pruning of vines. Note the 'blind' node on the cordon, lower right, which leaves a gap in the canopy at that point.

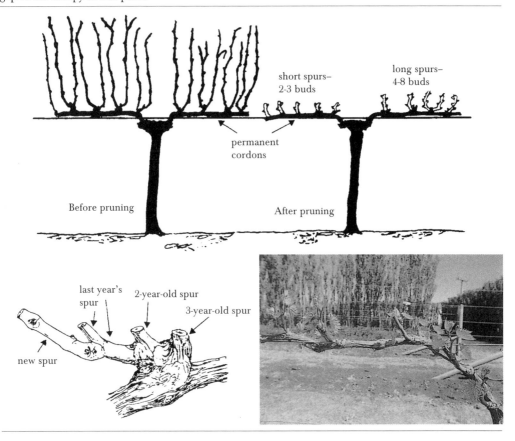

FIGURE 19: Cane pruning. The photograph shows vines with four canes selected and ready for tying down.

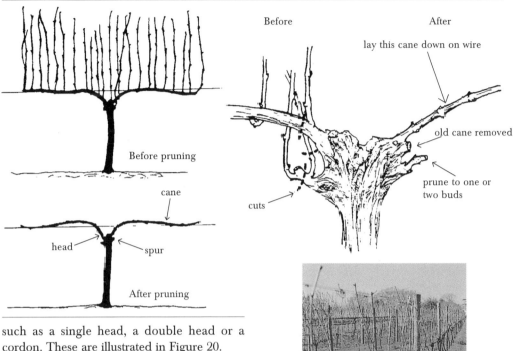

such as a single head, a double head or a cordon. These are illustrated in Figure 20.

Sometimes, canes are taken out on both sides, as illustrated in A, C, D of Figure 20, or may be placed only in one direction as indicated in Figure 20B. The advantages and disadvantages of each method follow.

- **HEAD CANE WITH CANES BOTH SIDES (FIGURE 20A)**

This common method of pruning allows growers to take one or two canes on either side, while leaving two spurs to provide replacement shoots for next season's canes. Although relatively simple to prune, the method creates problems of congestion near the head. This situation has three causes: *first*, the cane's basal buds are closer together than those further along the cane; *second*, the addition of spurs adds extra buds near the head; and *third*, buds on spurs tend to shoot and grow more vigorously than more distal buds.

There are ways to reduce these problems. For example, spurs can be dispensed with, but some growers worry that by doing this there may be inadequate replacement canes for next

Spur and Cane Pruning 27

FIGURE 20: Single head, canes both sides (A); single head, cane one side (B); double head (C); cordon cane (D).

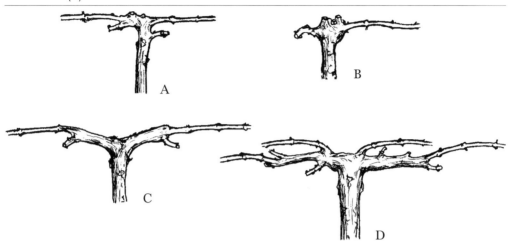

winter's pruning. In young plants, canes from the base of the previous season's canes are normally adequate. With older plants in less vigorous sites, dispensing with spurs may lead to replacement problems. If the head is below the wire on which the canes are tied so that shoots rise upwards from the head, then the Highest Point Principle (HPP) will somewhat compensate the TPP to reduce slightly the vigour of shoots near the head. Hand thinning of surplus young shoots near the head when they are less than 15 cm long can help prevent excessive congestion.

- **HEAD CANE WITH CANES ONE SIDE (FIGURE 20B)**

The head cane system described above is often used where plants are 1.4 m to 1.7 m apart in the row. For very close planting, say a metre or less (common for example in much of northern France), the head cane with canes on one side only is preferred. This is because a double-sided head cane in closely spaced rows must inevitably have short canes and therefore a higher proportion of basal buds. In addition, there are relatively more replacement spurs per metre of row. These problems are obviated if

FIGURE 21: Cane pruning variations for vines of different vigour and spacing. The cane and spur of the one-way variation on the lower left are alternated in position each year.

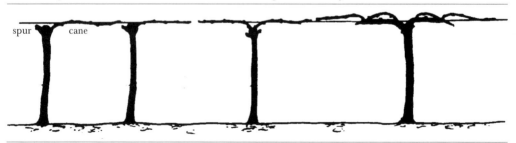

FIGURE 22: Positioning canes in the cordon cane system; below, cordon cane—small vine.

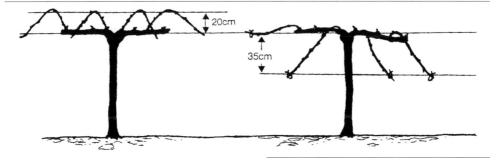

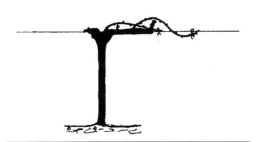

the canes point only in one direction. Each year the sides taken by canes or spurs are reversed.

- **THE DOUBLE HEAD (FIGURE 20C)**

The double head method has the trunk dividing about 30 cm below the base wire and each arm terminating in a head. The two heads are about 25 cm apart. This method avoids the congestion caused by a single head and allows space for replacement canes from the spurs to develop.

- **THE CORDON CANE (FIGURE 20D)**

This technique is a cross between the spur and the cane systems. A permanent cordon is retained that is shorter than the cordon used for spur pruning. Short canes originate from the cordon, and spurs are also retained to produce next season's fruiting canes. There are several features that may be beneficial in selected circumstances. For example, the geometry of the structure reduces congestion near the centre of the vine. Also, replacement spurs and canes may originate in different parts of the cordon and help reduce crowding. The method is particularly useful when wide spacing in the rows is used because it reduces the need to find extra-long canes to fill the gaps between vines. In vines closely spaced in the rows, a small cordon is sometimes left on one side, as shown in Figure 22.

Figures 21 to 22 show that canes may be trained along the base wire, or arched over a wire above the base wire, or tied down to a lower wire, or placed in various combinations of these configurations. It is possible to work out the merits of each of these alternatives using the principles outlined earlier in this chapter.

Figure 23 shows two techniques for securing a cane to a wire.

FIGURE 23: Tying cane to wire. The left-hand side of the diagram shows the use of a hand-held taping machine. The cane is not twisted around the wire, which makes removal easier next winter. The right-hand side shows tying at the end of a cane, which is twisted around the wire.

Spur and Cane Pruning

7 Vine Spacing

The number of vines per unit area is determined by the spaces between rows and between plants in the row. Spacing of vines affects light interception. For convenience, the discussion here is restricted to plants in rows on a VSP trellis.

High-density planting

High-density planting normally:
- reduces the plant size, both above and below the ground,
- utilises and exploits the soil volume more quickly and effectively,
- fills the above-ground area with leaves and fruit more rapidly and sometimes more thoroughly.

Of course, establishing high-density planting is more expensive because of the additional investment in plants and structures, but the rapid return to growers partly compensates for this.

High-density planting is appropriate when:
- the site is not too vigorous (an appropriate reduction in plant size will be hard to achieve in a vigorous situation),
- the site does not have limitations, such as a very low soil fertility or a likelihood of severe drought occurring in most summers (under such circumstances a greater volume of soil may be needed to provide adequate water and nutrients),
- appropriate machinery is available for adequate mechanisation of the operation,
- land is very expensive.

Note: exploitation of the soil volume is more complete the closer planting is to the square. For example, with plants spaced 2 m x 2 m, root-fresh weight has been found to be 87 percent greater than a rectangular planting of 3.50 m x 1.14 m, even though plant density per hectare is the same (Champagnol, 1984).

Low-density planting

Wide spacing is more appropriate:
- in vigorous situations, with vigorous varieties,
- where inadequate moisture is present for adequate growth,
- where narrow machinery is unavailable or too expensive,
- where the cost of land is not a major consideration.

Where the *row width* is predetermined by, for example, the availability of machinery, *plant distance in the row* will be close (say 1–1.5 m) if vigour of site and variety is low, and wide (1.5–2.5 m) if there is more vigour. Training systems may also influence planting distance; see later.

Some examples of high- and low-density planting are shown in Figure 24.

FIGURE 24: *Top*: close planting (1 m × 1 m) in Champagne; *middle*: moderately close planting in Napa, California; *bottom*: wide spacing in Barossa, South Australia.

8 Determining Bud Numbers

The more buds that are left after winter pruning, the greater the potential yield. Figure 25 indicates the yield response in relation to bud numbers. This generalised response curve shows that for yield, up to a certain point, a straight line relationship is observed: a given increase in bud numbers produces a predictable increase in yield.

This increase occurs because each bud that grows has approximately the same potential to produce a crop of berries. However, as more buds are retained, other factors come into play. For example, a large number of shoots may increase shade, which will reduce this year's crop because of EBSN and next year's because of poor inflorescence initiation. Furthermore, fewer buds are likely to grow if node numbers are high.

FIGURE 25: Relationship between bud numbers and yield and quality.

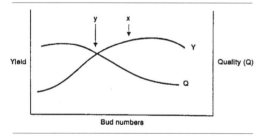

Ideally, growers should position their vineyards at x (Figure 25), the point of maximum yield without excessive bud numbers. However, the quality of juice may decline before point x is reached. The quality curve suggests that point y gives the best compromise between yield and quality.

The nature of the curve will vary according to several factors. However, as we note later, good canopy management can increase both the potential yield of the yield curve and delay the drop in quality of the quality curve, thus giving more fruit of a satisfactory quality standard.

Buds* may be conveniently expressed as either buds per hectare or buds per metre of row. Buds per metre of row can indicate the degree of congestion that may be expected. Smart and Robinson (1991) suggest that an ideal canopy has no more than 15 shoots per metre. Shoots closer than that are likely to provide too much shade for the lower leaves and bunches. Spacing of shoots at less than 15 per metre creates a situation where the vine crops below its potential. Divided canopies, as indicated later, may enable us to double the number of buds per metre of row while still maintaining shoot distance at the appropriate level.

The quotation of the number of buds per hectare should indicate the potential of a site to support a certain number of shoots without excessive stress or excessive crowding of these shoots. European viticulturists tend to use buds per hectare more than their New World counterparts. In classical European grape-growing districts, where growth-limiting soils are

---

* In this book we refer to 'buds per hectare' or 'buds per metre of row'. Some authors prefer the term 'nodes per hectare', this is because each node contains a compound bud that may give rise to one, two or even three shoots.

often chosen and irrigation is restricted or prohibited, it is important not to over-tax the shoot-growth potential of a site, a situation that can occur if bud numbers are too high; thus the value of 'buds per hectare'.

Classical and New World vineyards compared

In France, as previously noted, it is not uncommon for plants to be spaced at 1 m × 1 m. This spacing gives 10,000 plants per hectare, with each plant having one linear metre of row in which to develop. At 15 buds per metre, 150,000 buds are laid down per hectare. This bud number is greater than is common in many French Appellation vineyards, and we would therefore expect excellent light penetration to leaves and bunches.

In cool New World vineyards, rows are seldom this close, and 2.7 m is not an uncommon row spacing. At 15 buds per metre, the bud numbers would be 55,500 per hectare. Thus, with wide row spacing of 2.7 m compared with 1 m for narrow spacing, bud numbers are lowered almost to a third, if shoot density per linear metre is to remain the same. A low-yielding cultivar at 55,500 buds per hectare may yield only 5.5 t/ha, leaving growers tempted to increase bud numbers per metre, often to the detriment of quality and even yield.

Increased shoot numbers per metre above the 15 suggested may not always have negative effects on wine quality, and indeed many excellent wines are produced at numbers that easily exceed this. Ensuring adequate wine quality in such a canopy requires careful attention to *canopy trimming* and *leaf removal*.

On a vertical canopy, side trimming ensures that excess leaf density does not occur. Side trimming must be done regularly to avoid the situation where leaves and fruit alternate between dense shade before trimming and good exposure afterwards. Dense shade induces yellowing of leaves, which may not adequately photosynthesise even when, after trimming, they again receive improved light. Leaf removal at the fruiting zone at veraison ensures the final berry development occurs in a good light regime.

Magic numbers

The numbers of buds we leave per metre of row and/or per hectare obviously have a major influence on yield and quality. A key to successful viticulture is to find a yield that provides an economic return to the growers with no major loss of quality in the wine made from those grapes due to overcropping. The term *magic number** is the yield, in tonnes per hectare, which provides maximum return with optimum quality for a specified variety in a particular district. Over many years, growers and winemakers in Europe have determined their magic numbers and, in some cases, are so confident in these figures that maximum yields are specified by law.

We can achieve the same insight within our own vineyards by varying bud numbers and monitoring the yields and quality that are produced over a period of time. It may take several years to arrive at a satisfactory pruning level, but once found it can have considerable economic benefits to growers in later years. Their neighbours may also benefit from this knowledge, if it can be shared.

While such determination is very desirable, and should be encouraged, growers need some guidelines to know where to start. The following sections aim to give theoretical guidelines regarding expected outcomes from several pruning and training options. If we conclude

---

* **Magic Number** is another term I have coined to identify an important principle.

that for a specific variety and training system we should have a particular bud number per hectare or per metre of row, this figure can be the basis of our initial pruning strategy. At the same time, we should prune part of our vineyard with other bud numbers both above and below the theoretically determined figure. In time we may find one of these to be more appropriate to our needs. Without a theoretical guide to begin the quest, the search for the magic number will almost certainly be lengthier.

Estimating bud numbers

We need to begin by remembering that approximately 10 cm² of leaf area per 1 g of fruit is required. With this in mind, it can be shown that for every metre of row for a canopy about 30 cm wide and 1.5 m (150 cm) high, there is $(30 + 150 + 150) \times 100 = 33,000$ cm² of exposed leaf area. Thirty-three thousand cm² will therefore support $33,000 \div 10$, that is, 3,300 g of fruit (see Figure 26).

A high-yielding variety will average 350 g of fruit per shoot and need only $3,300 \div 350 = 9.2$ shoots per metre to provide an appropriate crop.

The equivalent figure for a medium-cropping variety (220 g) is $3,300 \div 220 = 15$ shoots per metre.

For a low-cropping variety (90 g), it is $3,300 \div 90 = 37$ shoots per metre. Here, growers would probably not use 37 shoots but may try, say, 25, which will still yield 2,250 g per metre.

Note: 3,300 g/metre is equivalent to 12.2 t/ha (2.7 m row width);
2,250 g/metre is equivalent to 8.3 t/ha (2.7 m row width).

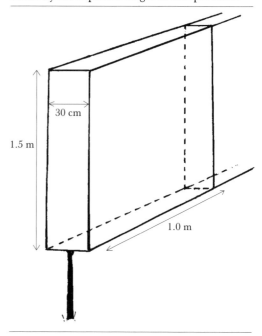

FIGURE 26: Dimensions used to estimate approximate leaf area. Leaf area per metre of row = $(150 + 150 + 30) \times 100 = 33,000$ cm². This calculation should satisfactorily ripen 3,300 g fruit. To determine the number of shoots per metre, divide by the expected weight of fruit per shoot.

9 Trellising

Many structures are used to support vines in world viticulture. Clearly, not all can be described in this small book. For this reason, we have had to make a choice and, rightly or wrongly, we describe those that we consider to be the most practical and widely used, along with some less used but potentially valuable techniques.

In some warm and sunny climates, vines are grown with minimum support. For example, the Goblet method of support, variations of which are seen in southern France, Australia and America, uses a small stake as support, but eventually the vine becomes self-supporting (Figure 27). The Goblet is used predominantly in hot areas with high light intensity and some limitation in vigour, such as summer drought. It is cheap to establish and manage, but it is not an appropriate system in cooler climates because vigour and congestion are very likely to lead to poor yields and quality.

In the Moselle Valley of Germany, the Hermitage area in France and other steep European vineyards, vines may be seen growing on poles. (We will describe this method later.)

FIGURE 27: Goblet-trained vines in Chateau Neuf du Pape, southern France.

FIGURE 28: Vertically shoot-positioned vines.

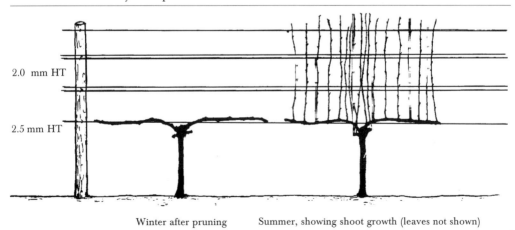

2.0 mm HT

2.5 mm HT

Winter after pruning   Summer, showing shoot growth (leaves not shown)

HT = high-tensile wire

These vines, together with vines using fruit trees as support (as occasionally seen in Italian vineyards), are unusual in not utilising posts and wires. Most trellises use wires.

The basic trellis has a base wire that supports the canes or cordons, with foliage wires helping to position the shoots in an appropriate configuration. The base wire normally takes more weight than the support wires and so has greater diameter – for example, 2.5 mm high-tensile steel.

Thinner and cheaper wires (such as 2.0 mm HT) are often appropriate for the foliage wires. Figure 28 indicates a not-unusual trellis with base and foliage wires.

The three general configurations encountered in vineyards, and illustrated in Figure 29, are:
- the single vertical plane,
- the double vertical plane,
- the non-vertical canopy.

FIGURE 29: Three configurations for vine trellises.

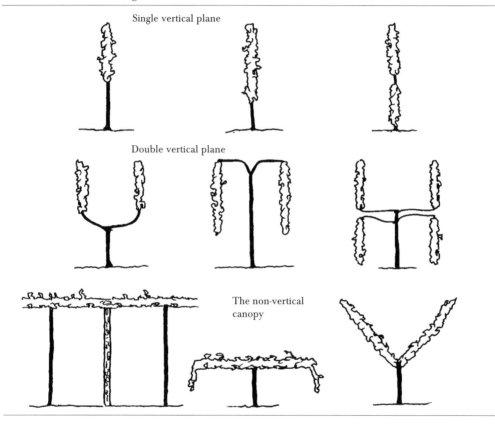

Trellises in the single vertical plane
THE SINGLE UPRIGHT
1: VERTICALLY SHOOT-POSITIONED (VSP)

The VSP is the system most commonly used in cool climates. The base wire is below the foliage wires, which are normally double, and contain the shoots within a narrow band above the base wire. The height of the structure varies from 1 m to 2 m, according to the distance apart of the rows (see Figure 30).

In the figure, A presents the situation for narrow row spacing, namely, 1–1.2 m. B shows that for wider spacing, namely, 2.7–2.9 m. They illustrate the two extremes. A is found in Champagne, Burgundy and other areas of France. Compared with B, A is less sturdy because the small plants put less strain on the trellis and its wires, posts are often metal stakes 5 m apart, and the wires tend to be strained less tightly. Trellises similar to that shown in B are common in many other places, such as Germany, Oregon (the United States) and New Zealand. The assembly must be much stronger because the height and weight of the vine can be considerable, especially when the vine is wet, the wind is strong or the soil is soft.

For the small trellis, shoots are trained between foliage wires by hand. Occasionally, the two wires are then clipped together to provide stability. Figure 31 illustrates an arrangement of wires for the large trellis. Here, the wires at (a) and (d) are fixed. Wire (a) is 2.5 mm high tensile. The rest are 2.0 mm. Wires at (b) and (c) can either be placed over the nails as shown at (1) or they can be hung before growth. As the canes

FIGURE 30:   Two versions of the VSP.

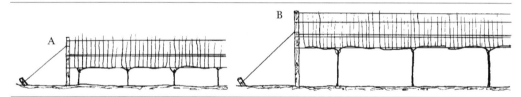

FIGURE 31:   End view along trellis, showing positioning of wires during shoot growth.

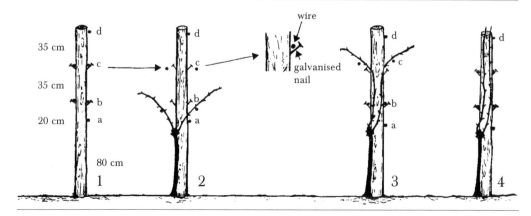

Trellising   37

grow, they fall over wires (b) and (c), as at (2) and (3), and are then lifted over the nails as shown at (3) and (4).

Another simple alternative uses a piece of wire, shown in Figure 32. When placed on either side of the post, this piece holds the wires in place at a convenient height. Wires may then be easily moved up and down so that in the early stages of vine growth they are close to the base wires. As shoots grow, the wires can be moved up towards the next set of double wires. This system gives early support to the growing shoots. Moving the wires upwards as shoots grow gives excellent and continuous support till the shoots reach the next wires. Figure 33 shows a commercial alternative for moveable wires. Remember that the base wire may be replaced by two wires 15–20 cm apart if double canes or arching are used (refer Figure 22, page 29).

Advantages of the VSP

This method of training is the most common in cool climates. Growers should therefore find it relatively easy to purchase structural materials, find expert help to erect it, and obtain advice on management of vines.

FIGURE 32: Attaching foliage wires to a post.

FIGURE 33: A commercial alternative to nails for supporting foliage wires, the 'Wire Care-Clip'.

The VSP readily adapts to mechanisation – trimming, harvesting and even mechanical pruning. Because the fruit is contained in a relatively narrow band along the row, sprays may be directed either to the whole canopy, if leaf and fruit cover is required, or just to the fruit zone if only fruit cover is needed. Other operations, such as leaf removal, are relatively simple.

Disadvantages of the VSP

The main problem with the simple upright is that achieving adequate yield in widely spaced rows requires a high-density laying down of buds. We have already considered this aspect and noted then that the problem is especially great with low-yielding varieties where, to get an adequate crop, up to 37 buds are needed per metre. This ratio is really too high, and growers must be satisfied with a lower figure and therefore a lower yield.

The VSP is more difficult to manage with varieties that do not naturally have strong upward-growing shoots.

Bud number recommendations

The data shown in Table 2 provide guidelines for bud numbers per metre. The table also takes into account varieties that have low, medium or high yields and shows the consequences of these two factors in terms of expected yields and leaf area per one gram of fruit produced.

Thus, growers wishing to produce a low-cropping variety on a 1 m × 1 m spacing should leave 14 buds per metre of row. This ratio gives 140,000 buds per ha and a yield in the order of 12.6 t/ha. This figure comes within the guidelines of less than 15 buds per metre and has over 10 cm² per one gram of fruit.

The figures from the wider spaced rows are less satisfactory for low-yielding varieties. These require 25 shoots per metre (higher than is ideal) but only yield 8.3 t/ha. The figures for middle- to heavy-cropping varieties on wide spacing are more appropriate – giving adequate yield (10 to 10.5 t/ha), suitably spaced shoots (8–13 buds/m) and sufficient leaf area (10 cm²/g).

Heavy-cropping varieties on narrow spacing, even at six shoots per metre, may only provide 7 cm² per gram of berry weight. In fact, however, such high-yielding varieties are seldom used in narrowly spaced vineyards. (See the following tables and also the insert 'Notes on Tables 2 to 8' on page 40 for additional information on this topic.)

Table summary. *With low-yielding varieties or when crops per metre of row are deliberately kept low, narrow spaced VSP seems ideal. For heavier-cropping varieties, wider spacing appears appropriate.*

Management

Winter pruning may be with spurs or canes. For *spurs*, a single cordon on both sides of the head is adequate; for *canes*, one or two on either side are used. If two canes are required, each may be tied to a separate wire about 15 cm apart, or they may be arched as indicated earlier (refer Figure 22, page 29). Cordon canes (Figures 20, 21 and 22, pages 28–29) are also used, especially if vines are widely spaced in rows. For close spacing in the row, canes may be positioned in only one direction (refer Figures 20 and 21, page 28).

TABLE 2:   Bud numbers for the VSP.

Type	Planting	Suggested buds per metre of row			Buds per hectare (Yield in tonnes/ha)			Exposed leaf area per hectare in m² (Leaf area per 1 g fruit in cm²)		
		L*	M	H	L	M	H	L	M	H
Narrow	1 m × 1 m⁺	14	9	6	140,000	90,000	60,000	15,000		
					(12.6)	(15.4)	(21.0)	(11.9)	(9.7)	(7.1)
Wide	2.7 ×	25	13	8	92,500	48,000	29,500	10,500		
	(1.2–1.5 m)^Φ				(8.3)	(10.5)	(10.4)	(12.1)	(10.2)	(10.4)

Key *L = low-yielding variety 90 g/shoot average; M = moderate-yielding variety 220 g/shoot average;
H = high-yielding variety 350 g/shoot average.
⁺ = Height of foliage 0.60 m.
Φ = Height of foliage 1.5 m.

Shoots that grow in the spring need to be positioned between support wires. The sooner this happens the better because a more satisfactory narrow canopy is achieved, that is, one that has less volume and less internal shading.

Capfall normally occurs when 16 to 18 leaves have formed on the shoot. At this stage they will be at the top wire or even beyond and will be almost ready for topping. (Topping a few days before capfall may increase fruit set but, unfortunately, early topping encourages lateral shoot growth and, in some varieties, a second crop on these laterals.)

Later in the season, usually after veraison, growers may consider leaf plucking around the bunches in dense canopies.

Vigour rating

In an ideal environment, shoots would grow to about 17 leaves, at which stage extension growth should slow down and stop. In most cases, the ideal is not achieved, and summer trimming or topping is required. The number of toppings and trimmings needed to maintain a well-exposed, not excessively shaded canopy will indicate the vigour of the site. If more than two are required, we might assume that vigour is somewhat excessive and that a divided canopy is more appropriate. The single upright can be relatively easily converted to a divided upright, such as the Scott-Henry, which we shall describe shortly.

Narrow spacing is the least appropriate for a vigorous site, and best for one with a low to moderate vigour rating. Wider spacing is suited to moderate, but not excessive, vigour.

Notes on Tables 2 to 8

Tables 2 to 8 (pages 39–54) aim to provide growers with an overview of the advantages and disadvantages of various training systems. They also provide some of the constraints and limitations of each system imposed by the fixed variables of the trellis system and the number of buds that are left. Figure 34 should help to put this explanation in perspective.

The trellis configuration and the distance apart of the rows determine the exposed leaf area in the vineyard. Because the trellis is unlikely to be modified, we call it a fixed variable, and because the exposed leaf area is dependent on the structure, we call it a dependent variable. The yield to be expected is dependent on (1) the leaf area, (2) another fixed variable – the cropping potential of the variety we have planted, and (3) the bud number, which can be varied each year and is termed a variable input. The computations on pages 39 to 54 allow us to estimate the leaf area per gram of fruit.

The tables show the consequences of changing some of the variable inputs. For example, if the expected yield or the leaf area per gram of fruit is not what we think is desirable, we can change the bud numbers till they provide a more acceptable figure, as is indicated by the upwards-pointing arrows on the right of the diagram. If the figures are still not appropriate, we can change the more fixed variables, shown by the arrows on the left-hand side – easy to do if we have not planted; difficult, but not necessarily impossible, once planted.

FIGURE 34: Dependent and fixed variables and vine yields.

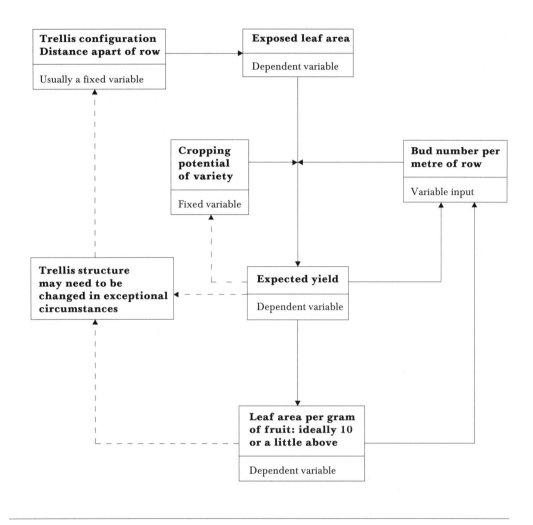

THE SINGLE UPRIGHT
2: USING DOWNWARD-POINTING SHOOTS

The 'Curtain' or High Sylvoz

This system is a little like the VSP in reverse. The main cordon wire is at the top of a 1.7 m trellis, with shoots hanging down rather than being trained upwards (Figure 35a, front cover).

Short canes plus spurs are selected, preferably with the canes pointing in a downwards direction. Alternatively, canes can be loosely tied on wires on either side of the head (ie, no permanent cordon). Shoots that develop in spring are encouraged to hang downwards, a process that growers can aid by having two foliage wires about 50 cm below the cordon.

Advantages of the Curtain

The Curtain is the simplest and one of the cheapest ways of training vines. Apart from encouraging shoots downwards, growers need not position shoots. In winter, they need not drag out canes from between the wires but can simply cut them and let them fall on the ground. Fruit exposure is good. Exposure of buds at the base of canes to enhance next season's crop is likewise very good.

The method is easily adapted for mechanical harvesting. Shoots, if they grow 1.7 m to the ground, provide a leaf area which can theoretically support a higher crop than the single upright (see calculations for bud numbers). Because there is less need to top the shoots, fewer laterals form and fewer second-crop berries occur. The extra height above the ground sometimes provides protection from a frost layer below.

The system is good for varieties that are non-upright in growth habit.

Disadvantages of the Curtain

Foliage management, although not time-consuming, must be undertaken carefully. If shoots are not positioned downwards before fruit set, berries will develop in shade and, on later exposure, will very likely sunburn. Despite this, well-exposed fruit in high light intensity regions, such as Australia and New Zealand, may still sunburn; even mild sunburn may

---

FIGURE 35a: The Curtain or High Sylvoz.

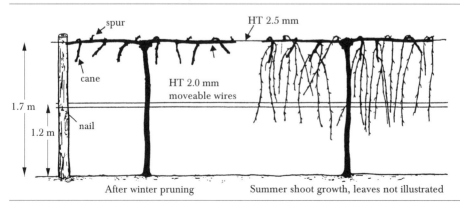

After winter pruning   Summer shoot growth, leaves not illustrated

exacerbate splitting and disease. In addition, berries may produce excess tannin, giving wine a bitter characteristic. Bird damage may sometimes be greater with this system.

Because the system is so easy to manage, growers are sometimes tempted to overcrop, a practice that leads to low °Brix and poor quality wine. One other disadvantage is that, because the trunk is up to twice the height of normal, the first crop may be delayed by one year. Growers occasionally crop at mid-height first and take the trunk to the top wire the next year.

Bud number recommendations

Bud recommendations are similar to the VSP on wide spacing, although the bigger leaf canopy allows us to increase buds per metre and achieve slightly higher yields.

Table summary*. This system, like the VSP at wider spacing, is appropriate for medium- to heavy-cropping varieties. Low-yielding grapes even at 25 buds per metre do not give a high return.*

Management

About two weeks before capfall, growers should encourage shoots to fall downwards, a process that may be assisted by having the training wires at 1.2 m above the ground. This situation is the reverse of the VSP, where wires pull the shoots upwards. Normally, once positioned, no further training is required, although some growers prefer to trim the shoots above the ground to encourage air circulation. Despite the grower's best efforts to position wires downwards, some shoots will grow above the wire. If few in number, they may be left to provide a little shade to the exposed bunches. Otherwise, they may be pulled off by hand or trimmed about 25 cm above the cordon.

Vigour rating

Compared with the VSP, the Curtain provides more space for shoot growth without trimming and so is rather easier to manage on a slightly more vigorous site. As indicated, other things being equal, this extra leaf area can ripen a slightly heavier crop. If growth of shoots continues vigorously after shoots reach the ground, canopy division may be advisable. The Geneva Double Curtain, described below, is the appropriate next step, if there is adequate distance between rows.

TABLE 3: Bud numbers for the Curtain.

Type	Planting	Suggested buds per metre of row			Buds per hectare (Yield in tonnes/ha)			Exposed leaf area per hectare in m^2 (Leaf area per 1 g fruit in cm^2)		
		*L	M	H	L	M	H	L	M	H
Curtain	2.8 × 1.5$^+$	25	17	11	89,000	60,500	39,000		13,000	
					(8.0)	(13.4)	(13.7)	(16.5)	(9.9)	(9.6)

Key *L = low-yielding variety 90 g/shoot average; M = moderate-yielding variety 220 g/shoot average;
  H = high-yielding variety 350 g/shoot average.
   $^+$ = Height of canopy taken as 1.7 m.

THE SINGLE UPRIGHT
3: USING UPWARD- AND DOWNWARD-POINTING SHOOTS

Mid-height Sylvoz: (Version A Figure 35b)

Pruning for this method uses the cordon cane with only downward-pointing canes (refer Figure 22, page 29; Figure 35b). Upward shoots form from the spurs on the top and are trained between the foliage wires. Shoots from downward-pointing canes are trained downwards in a manner not dissimilar to the Curtain.

Advantages of the Mid-height Sylvoz

Because some shoots are trained upwards and others downwards, this method acts like a divided canopy. The division reduces congestion in the area of origin and gives greater possibility of achieving an appropriate shoot density. Light penetration into the canopy should be good.

Disadvantages of the Mid-height Sylvoz

Vines trained to this system look rather similar to the Scott-Henry, shortly to be described. It is probably a little easier to manage, but given that not so many fruiting shoots point upwards, there is rather more congestion in the lower half than the upper. Because some downward-pointing shoots are closer to the ground than others, fruits will be subtended by different lengths of shoots and therefore different numbers of leaves. There will probably be slightly more between-bunch variation than in more ordered systems. The fruit zone is quite widely distributed, which is good for light penetration, but slightly more spray will be required for 'fruit only' applications.

Bud number recommendations

The Mid-height Sylvoz has many elements of a divided canopy because replacement shoots on spurs are separated from the main fruiting shoots on canes. The effective bud number per metre of canopy will therefore be half the buds per metre of row.

Management

The upward-pointing shoots may be used as next year's canes. The low congestion at the

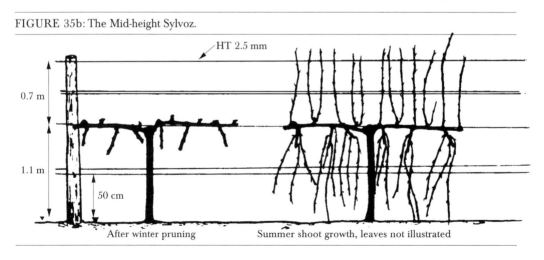

FIGURE 35b: The Mid-height Sylvoz.

top means that the buds should be very productive. Summer topping and trimming may be needed to contain the growth of the upward-pointing shoots. The downward-pointing shoots may also need trimming to keep them off the ground and allow adequate air circulation. Some growers do not bother to position the downward-pointing shoots that grow outwards from the rows and they are simply trimmed off with lateral trimming operations. The possible disadvantage of this practice is that those berries subtended by severed shoots may be of lower ripeness.

Vigour rating

This is an appropriate system for vineyards of moderate vigour.

Mid-height Sylvoz (Version B)

The main difference between this and the previous situation is that some of the shoots from downward-pointing canes are allowed to grow upwards. Ideally, this practice equalises the number of upward- and downward-directed shoots. This alternative should give slightly better bunch exposure on the lower portion of the canopy but slightly less fruitful buds on canes to be used next year. It has many similarities to the Scott-Henry.

FIGURE 36: Mid-height Sylvoz with four canes tied to a wire below the cordon.

Mid-height Sylvoz (other versions)

Instead of the larger number of shorter, downward-pointing canes shown in Figure 35b, growers often reduce the number to four per plant (Figure 36), which gives a wider distribution of the fruiting zone.

Table summary. *Even low-cropping varieties can give fair yields (10 t/ha) on the Sylvoz. Shoot density per metre of row is low for moderate-to high-yielding grapes, yet yields are very satisfactory (14 t/ha).*

TABLE 4: Bud numbers for the Mid-height Sylvoz.

Type	Planting	Suggested buds per metre of row (Buds per metre/canopy)			Buds per hectare (Yield in tonnes/ha)			Exposed leaf area per hectare in m² (Leaf area per 1 g fruit in cm²)		
		*L	M	H	L	M	H	L	M	H
Mid-height Sylvoz	2.7 × (1.2–1.5)⁺	30 (15.0)	18 (9)	11 (5.5)	111,000 (10.0)	66,500 (14.7)	40,500 (14.3)	13,500 (13.7)	(9.3)	(9.6)

Key *L = low-yielding variety 90 g/shoot average; M = moderate-yielding variety 220 g/shoot average;
    H = high-yielding variety 350 g/shoot average.
    ⁺ = Height of canopy taken as 1.7 m.

The Scott-Henry

This system (shown in Figure 37) divides the canopy on a single upright trellis in a slightly more formal way than the Mid-height Sylvoz. Cane pruning is currently used for the Scott-Henry, but spur pruning is a possibility worth investigating. If this latter method is used, Smart and Robinson (1991) suggest using different plants to supply cordons for the low and high wires.

Advantages of the Scott-Henry

The Scott-Henry is well adapted to sites where the vigour is moderate to high because it allows growers to lay down more buds without excessive shoot density. While rather more attention is needed than with the VSP upright, the work load is not excessive. Unlike the Mid-height Sylvoz, the shoots have similar length, and growers therefore can expect less variation in ripeness. Vines may be mechanically harvested. Good exposure of shoots may not only improve quality but also increase fruitfulness for next year's canes.

Disadvantages of the Scott-Henry

A key factor with the Scott-Henry is that the lower downward-pointing shoots must be positioned at the correct time. Failure to do this may mean excessive congestion because of the lower shoots growing into the upper ones. The slightly wider distribution of bunches means a little more spray is needed for fruit-only applications (but because of reduced canopy density, penetration might be better). Growers sometimes complain of lower ripeness levels in the bunches carried by downward-pointing shoots, a situation brought about by shading from adjacent rows and accentuated if the row distance is too narrow.

Bud number recommendations

It can be seen from the figures in Table 5 that the Scott-Henry has a potential close to the narrow-spaced VSP in terms of (i) a low number of shoots per metre of canopy and (ii) a satisfactory expected yield.

Good separation of upward- and downward-pointing shoots in Scott-Henry.

FIGURE 37: The Scott-Henry Training System.

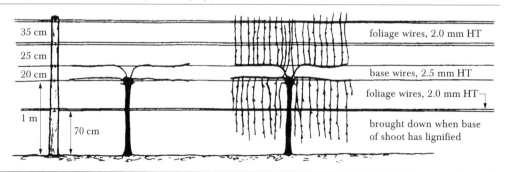

46 *Monographs in Cool Climate Viticulture – 1: Pruning and Training*

TABLE 5: Bud numbers for the Scott-Henry.

Type	Planting	Suggested buds per metre of row (Buds per metre/canopy)			Buds per hectare (Yield in tonnes/ha)			Exposed leaf area per hectare in m² (Leaf area per 1 g fruit in cm²)		
		*L	M	H	L	M	H	L	M	H
Scott-Henry	2.7 × (1.2–1.5)⁺	30 (15.0)	18 (9)	11 (5.5)	111,000 (10.0)	66,500 (14.7)	40,500 (14.3)	13,500 (13.7)	(9.3)	(9.6)

Key *L = low-yielding variety 90 g/shoot average; M = moderate-yielding variety 220 g/shoot average; H = high-yielding variety 350 g/shoot average.
⁺ = Height of canopy taken as 1.7 m.

Table summary. *Like the Mid-height Sylvoz, low-cropping varieties give fair yields of 10 t/ha. Medium to high producing grapes give good yields (14 t/ha) with low shoot density and adequate leaf/fruit ratio.*

Management

We have already emphasised the importance of shoot positioning, particularly of the lower downward-pointing shoots. The upper shoots are trained between wires in a manner similar to the VSP. The lower shoots are moved away from the upright by easing out from the wires above, beginning when shoots are about 20 cm long and repeating once or twice until about three weeks before capfall. At this time, the shoots are fixed in position with the moveable wire below, a practice that is easier if the side on which the lower shoots are trained is the predominantly leeward side.

The upper shoots are topped as for the VSP. The lower ones tend to grow less vigorously but may be trimmed before they reach the ground for better air circulation.

Vigour rating

The Scott-Henry is appropriate for vineyards of moderate to high vigour. If, after adopting the Scott-Henry, growers need to carry out more than two trimmings, then it is likely that the grapes are growing on the wrong site. Some relief may be gained by halving the number of plants (that is, removing every second plant) to achieve wide spacing between vines. If this is done, spur or cordon cane pruning will be needed. Other methods to control excess vigour are discussed in Chapter 10 – Trouble Shooting.

Trellises in the double vertical plane
THE GENEVA DOUBLE CURTAIN (GDC)

Developed by Dr Nelson Shaulis of New York State to cope with vigour found in widely spaced rows, the GDC has been used since the 1960s (Shaulis *et al*, 1966). Basically, it is a double version of the Single Curtain previously described, and during the growing season appears as shown in Figure 38.

Because the parallel hanging curtains are 1–1.5 m apart, the rows themselves also need to be more widely spaced, probably about 3–3.5 m. There are several versions of the GDC, but that shown in Figure 38 is probably the best. Vines will be spaced at 1.5 m to 2.5 m depending on vigour (the most vigorous will, of course, be at wider spacing), and each vine may be trained either on both sides or on one side (see Figure 38 below).

Advantages of the GDC

The advantages claimed for the Single Curtain, such as simplicity of pruning and training, also apply to the GDC. Although the rows of the Double Curtain will need to be further apart than the Single Curtain, the extra curtain increases the total effective curtain length by a factor of about 50 percent at very little loss of light interception. Thus, increased yields can be expected. (It is not a 100 percent increase because the rows are wider apart than the Single

FIGURE 38: The Geneva Double Curtain.

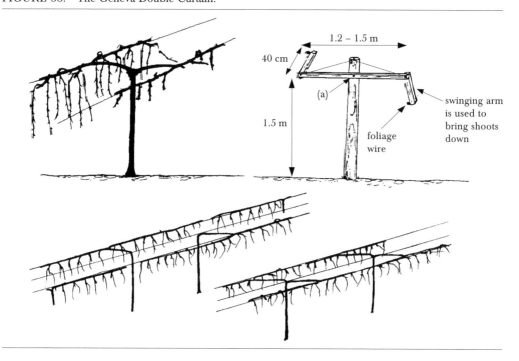

Curtain.) The system can be mechanically harvested, but less damage will occur if, instead of the cross-arm being rigid, a pivot is placed at point (a) in Figure 38. The pivot allows upward movement of the whole canopy while the grower is harvesting.

Disadvantages of the GDC

Timely attention is essential with this sytem, even more so than with the Single Curtain. The common mistake of allowing foliage to fill the gap between the two curtains creates a microclimate that completely annuls any advantages of the split canopy. As with the Single Curtain, growers need to be aware of sunburn problems. Two parallel planes pose some problems for spraying because, with both sides of the canopy not being adjacent to the nozzles, higher pressures are needed to reach the centre.

Bud number recommendations

Table 6 shows the potential of the GDC to accommodate high bud numbers and give high yields while still allowing good shoot spacing and appropriate leaf/fruit ratio. Note, however, that the effective leaf area is probably over-estimated in the GDC given that the lower inside leaves will not be fully effective because of reduced light at that point.

Table summary. *The GDC gives yields, within the appropriate parameters of bud numbers per metre of row and leaf/fruit ratio, that are as good as the narrow VSP. Even with low-yielding varieties, crops of 12 t/ha are obtained and for higher yielding varieties 20 t/ha are expected.*

Management

We have already stressed the importance of foliage management. The approach taken for single curtains should be adopted for the GDC. The pivoting fruiting wire shown in Figure 38 (page 48) will assist this operation.

Vigour rating

The extra loading that can be achieved without loss of leaf and fruit exposure means the system is good for moderate to vigorous situations.

TABLE 6: Bud numbers for the Geneva Double Curtain.

Type	Planting	Suggested buds per metre of row (Buds per metre/canopy)			Buds per hectare (Yield in tonnes/ha)			Exposed leaf area per hectare in m² (Leaf area per 1 g fruit in cm²)		
		*L	M	H	L	M	H	L	M	H
Geneva Double Curtain	3.3 ×1.8⁺	44 (22)	30 (15)	20 (10)	133,500 (12.0)	91,000 (20.0)	60,500 (21.2)	22,500 (18.7)	(11.2)	(10.6)

Key *L = low-yielding variety 90 g/shoot average; M = moderate-yielding variety 220 g/shoot average;
H = high-yielding variety 350 g/shoot average.
⁺ = Height of canopy 1.7 m.

THE LYRE

Developed by Dr Carbonneau in Bordeaux, the Lyre is similar in principle to the GDC, but the shoots point upwards rather than downwards. Carbonneau (1979) believes that the quality achieved by very narrow row spacing in much of France might be achieved more cheaply and as effectively by such a divided canopy. His results suggest that adoption of the Lyre may, in fact, do this. However, in the Appellation districts of France, the authorities do not allow any departure from the system normally used. Figures 39 and 40 and the front book cover show the structure and basic details of training.

accessible for training or sprays. If the centre of the U is filled by leaves and shoots due to poor management, all advantages of the system are lost. Therefore, as with the VSP, trimming and topping is usually required, and a slightly more

FIGURE 39: The Lyre.

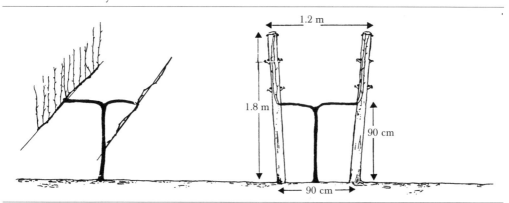

FIGURE 40: A metal structure designed for the Lyre. The arms can be moved towards the horizontal to increase light interception when inter-row cultivation, spraying and the like are not being applied.

Advantages of the Lyre

Each side of the Lyre is essentially a VSP. Having two VSPs in the same structure achieves an effect similar to having twice the number of separate narrow rows. Because the fruit is near the base, it receives less light than fruit in the GDC. However, the problems of sunburn that are sometimes noted for the Double Curtain are lessened.

Disadvantages of the Lyre

Any system with parallel canopies poses the problem that the inside area is not readily

50   *Monographs in Cool Climate Viticulture – 1: Pruning and Training*

TABLE 7: Bud numbers for the Lyre.

Type	Planting	Suggested buds per metre of row (Buds per metre/canopy)			Buds per hectare (Yield in tonnes/ha)			Exposed leaf area per hectare in m² (Leaf area per 1 g fruit in cm²)		
		*L	M	H	L	M	H	L	M	H
Lyre	3.2 × (1.2–1.5)⁺	36 (18)	26 (13)	16 (8)	112,500 (10.1)	81,000 (17.9)	50,000 (17.5)	(16.7)	17,000 (9.4)	(9.6)

Key *L = low-yielding variety 90 g/shoot average; M = moderate-yielding variety 220 g/shoot average;
   H = high-yielding variety 350 g/shoot average.
   ⁺ = Height of canopy 1.2 m.

complicated mechanical cutting system will be needed for the Lyre than the single VSP. The Lyre does not adapt to standard mechanical harvesters, and a specially designed one is therefore being produced in France.

Bud number recommendation

Yields and leaf/fruit ratio for the Lyre (Table 7) are not dissimilar to the narrow spacing VSP that it is intended to replace. Shoot density is a little higher, but not excessively so.

Table summary. *Figures for the Lyre look good and, as with the narrow VSP and the GDC, satisfactory crops (10 t/ha) for low-yielding varieties are obtained. Crops of 17–18 t/ha can be expected from medium- to high-yielding grapes.*

Management

Growers manage each canopy in a manner similar to the single VSP, using moveable foliage wires, along with trimming and topping. Earlier versions of the Lyre had a rather more narrow base, but this was found to cause excess shading near the fruit zone, which led to the recommendation to use two vertical or near vertical canopies (refer Figures 39 and 40, page 50).

Vigour rating

Because the parallel canopies on the Lyre have a lesser volume for shoots and leaves than the GDC, they are suited to a less vigorous site, possibly similar to that for the Scott-Henry.

THE RUAKURA TWIN TWO TIER

FIGURE 41: The Ruakura Twin Two Tier.

The principle of this system (Figure 41) is similar to the Lyre, with two exceptions: the space between the two parallel canopies is greater (1.5 m compared to approximately 1 m), and each canopy is split vertically like the Scott-Henry. Normally, the upper and lower tiers come from different plants.

Rows are set at a minimum of 3.6 m wide and plants within rows at approximately 2 m apart. The fact that a single plant covers a fair area of trellis means that vigour control, especially in the early years, is good and that the system is suited to a vigorous site.

Smart and Robinson (1991) provide a full discussion of this system, so rather than giving further details here, we refer readers to that publication. Growers could probably achieve similar results to the Ruakura with a Scott-Henry that has between-row spacing alternating between 1.5 m and 2.1 m.

Growing on a pole

In cool areas, where every degree of heat counts, growers often use steep slopes facing the sun. If these slopes are not terraced, trellises are difficult to erect and poles can be more suitable (Figure 42). Tractors may be difficult or impossible to use, and extensive hand labour may be required. For this reason, every piece of ground is used, and the poles are often very close together — sometimes little more than one metre apart each way. A modified head cane is used for plants grown on a pole (see Figure 42 below).

FIGURE 42: Training on a pole. The photograph shows vines on poles in the Moselle district of France.

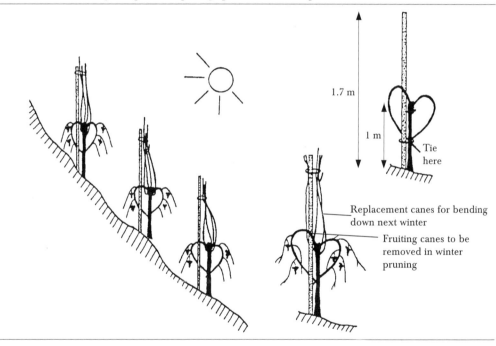

Trellising

Making a decision

Table 8 summarises the bud number recommendations presented individually for most training systems. The data in this table give guidelines that growers might follow in the early years but modify as they gain experience and an understanding of their own vineyards.

The key that follows may help growers choose a system. Remember that the choice within each group will depend on a number of factors, many of which have been mentioned earlier in this chapter.

TABLE 8: Bud number for all training systems.

Type	Planting	Suggested buds per metre of row (Buds per metre/canopy)			Buds per hectare (Yield in tonnes/ha)			Exposed leaf area per hectare in m² (Leaf area per 1 g fruit in cm²)		
		*L	M	H	L	M	H	L	M	H
Narrow VSP	1 × 1 m	14 (14)	9 (9)	6 (6)	140,000 (12.6)	90,000 (15.4)	60,000 (21.0)	(11.9)	15,000 (9.7)	(7.1)
Wide VSP	2.7 × (1.2–1.5) m	25 (25)	13 (13)	8 (8)	92,500 (8.3)	48,000 (10.5)	29,500 (10.4)	(12.1)	10,500 (10.2)	(10.4)
Curtain	2.8 × 1.5 m	25 (25)	17 (17)	11 (11)	89,000 (8.0)	60,500 (13.4)	39,000 (13.7)	(16.5)	13,000 (9.9)	(9.6)
Mid-height Sylvoz	2.7 × (1.2–1.5) m	30 (15)	18 (9)	11 (5.5)	111,000 (10.0)	66,500 (14.7)	40,500 (14.3)	(13.7)	13,500 (9.3)	(9.6)
Scott-Henry	2.7 × (1.2–1.5) m	30 (15)	18 (9)	11 (5.5)	111,000 (10.0)	66,500 (14.7)	40,500 (14.3)	(13.7)	13,500 (9.3)	(9.6)
Geneva Double Curtain	3.3 × 1.8 m	44 (22)	30 (15)	20 (10)	133,500 (12.0)	91,000 (20.0)	60,500 (21.2)	(18.7)	22,500 (11.2)	(10.6)
Lyre	3.2 × (1.2–1.5)⁺	36 (18)	26 (13)	16 (8)	112,500 (10.1)	81,000 (17.9)	50,000 (17.5)	(16.7)	17,000 (9.4)	(9.6)

Key *L = low-yielding variety 90 g/shoot average; M = moderate-yielding variety 220 g/shoot average;
H = high-yielding variety 350 g/shoot average.
⁺ = Height of canopy 1.2 m.

- Vigorous site, *consider*:
    Geneva Double Curtain
    Scott-Henry
    Single Curtain or Sylvoz
    Ruakura Twin Two Tier
- Moderately vigorous site:
    Scott-Henry
    Single Curtain
    Sylvoz
    Lyre
- Low to moderate vigour site:
    Scott-Henry
    Sylvoz
    Lyre
    VSP wide spacing
- Low vigour site:
    VSP wide spacing
    VSP narrow spacing.

10 Trouble Shooting

Problems that arise in the vineyard may be related to the method of training used. This chapter considers the more common problems and possible ways to reduce them.

Shoots variable in length and vigour

This problem (illustrated in Figure 43) tends to be more common in vineyards that have been cane pruned. The following variations may be found:
i) *Vigorous shoots containing few or no flowers*: These often come from old wood on trunk or cordons.
ii) *Vigorous shoots containing a full complement of bunches*: These are more common at the ends of canes (End Point Principle) or at other favoured positions.
iii) *Less vigorous shoots, possibly with fewer or smaller bunches*: These are usually found at less-favoured positions on the canes.
iv) *Small, weak shoots with one to three leaves but a full complement of bunches*: These may be found at all positions on the cane, but are more common on less-favoured ones.
v) *Buds that do not burst and remain dormant (called 'blind buds')*: These may be dead buds or the plant may have too little vigour to induce them to grow.

Two major factors probably contribute to these variations. First, the vine may have been under-pruned and the plant simply does not have the energy to make all the buds grow to the same level; (iv) and (v) may be the consequences of such under-pruning. The second cause may have been excessive shading of the shoots in the previous growing season, as with (iii). Under these conditions, dead buds may be common, as with (v). If the vine is over-pruned, most buds growing from canes will be of the type (ii).

FIGURE 43: Shoot variation in Riesling.

Shoots (i) will also appear from old wood on the vine.

The disadvantage of excessive vigour variation is that this situation may also induce variation in speed of ripening. There is evidence that grapes on shoots with fewer leaves ripen later and often have reduced sugar levels. Few winemakers welcome this variation in ripeness. Variation in vigour of shoots along the cane also contributes to differences in inflorescence initiation.

Growers can use two approaches to alleviate this problem. The first is to consider spur pruning of vines. Variation in vigour, as we have already noted, is less in spur-pruned vines. The second approach is to reassess the severity of pruning.

Uneven shoot distribution

The desirability of having shoots distributed as evenly as possible along the row has been emphasised in this monograph. Variability in shoot vigour makes this more difficult, but another factor can also contribute to unevenness.

Tables 2 to 8 indicate the importance of varying bud numbers per metre of row to achieve desirable cropping levels. If a vineyard is spur pruned this can be achieved by modifying the number of buds on a spur without causing uneven shoot distribution. With cane pruning, however, it can be difficult to adjust cane number after pruning to give exactly the right number, as indicated earlier. Thus, if canes are 1 m long and have on average 10 buds each, it will be easy to get 10 buds per metre, or by doubling canes, 20 buds; to get 15 is not so simple.

Growers may increase cane length to 15 buds and overlap them at the end. This, unfortunately, doubles the shoot density at exactly the point where vigour is most extreme due to the End Point Principle. A second congested area occurs at the head due to the trunk proximity principle and any extra spurs retained. The space between is less dense and of lower vigour. Options that alleviate the problem are not many, but if canes are left sufficiently long so they overlap with their ends coinciding with the middle of the cane from the neighbouring vine, a slightly better distribution occurs. Alternatively, improvement may be found if a second shorter cane is used to increase bud numbers. The length should be such that the end point of the short cane coincides with the middle of the first cane.

Growers who feel it necessary to cane prune may consider the cordon-cane as a compromise. This reduces the cane length and evens out the variability along the canopy.

Vines always too vigorous

If a grower suspects that the vigour in the vineyard is too high there are two methods, using cane weight measured after winter pruning, that can be used to confirm this. See Smart and Robinson (1991) for a more detailed discussion.

Pruning weight measurements are conveniently expressed as either kilograms per metre of row length, or mean cane pruning weight.

- *Kilograms per metre*: If pruning weight exceeds about 0.6 kg/m row length, then the canopy is vigorous and likely to be too shaded.
- *Mean cane pruning weight*: If the mean cane weight is below 20 g, then the vines are excessively devigorated (by stress or by too light pruning). If more than about 40 g, then the vines are likely to be out of balance, with too high a leaf to fruit ratio. As a general rule, we should be able to prune to about 30 buds per kilogram of pruning weight.

In discussing various training systems, we have stressed those that might be appropriate for more, or less, vigorous situations. If the problem is not easily rectified by, say, canopy division, the following techniques may help.

TRUNK GIRDLING

Girdling is sometimes used on less-fruitful cultivars to increase fruit set. Treatment is recommended at about 10 days before capfall when fruit set may increase but, in addition, some vigour reduction may occur. It is also possible to girdle at other times of the year with these general results:

- Girdling from when shoots are 20 cm long until approximately *10 days before capfall*: The earlier girdling is applied, the greater will be the vigour reduction and the less the effect on fruit set.

- *Girdling 5 to 10 days before capfall*: As discussed above, this method results in improved set and reduced vigour.
- *Girdling at veraison*: This practice may advance ripening but will probably have less effects on vigour reduction. However, growers may notice some carry-over to next year.

Growers should experiment with girdling in their own vineyards, testing a few rows rather than the whole vineyard. We have girdled vines in three consecutive years with no apparent deleterious effects on the vines. Girdles are done with a double-bladed parallel knife down to the wood. Removing loose bark before girdling makes the operation simpler. The width of the girdle should be no more than 3 mm, and it should reach to the wood (xylem) but not penetrate it. The girdle should go completely around the trunk (Figure 44). We do not recommend girdling weak vines.

Effective use has also been made of other approaches. For example, a single cut with a pruning saw no more than 3 mm wide through the bark around the trunk will achieve similar results. Instead of girdling the trunk, canes can be girdled around their base using 'pliers' as shown in Figure 44.

CHEMICALS

Some growth inhibitors will reduce vigour. Chlormequat (CCC, 'Cycocel'), at approximately 300 ppm, has been used to improve set and reduce internode elongation, but it is not permitted in many viticultural areas. PP333 ('Cultar') also reduces vigour, but useful concentrations and methods of application have yet to be determined.

Ethephon ('Ethrel') applied during the growing season terminates growth of shoots for a period of up to six weeks. Three hundred to 800 mg/litre (ppm) can be evaluated. ('Ethrel'

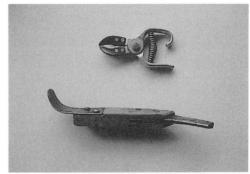

FIGURE 44: *Above*: tools used for girdling – canes at top; trunk at bottom. *Below*: Girdled trunk, with last year's girdle at top.

contains 48 percent ethephon; concentrations just given are of the active ingredient – ethephon.) It should *not* be applied before three weeks after capfall otherwise cropping will be decreased. In addition, spray is best directed at all times to the shoots and not the berries. Applied at or after veraison, spraying may advance ripening. Check locally for the legality of using ethephon and, if using, test on a portion of the vineyard in the first year to assess effectiveness for local conditions.

CONTROL BY CROPPING

A heavy-yielding vine tends to have less vigour, and shoots with large bunches terminate their growth much earlier – often at 15 to 20 nodes. The vigour of the site will determine the weight of crop per shoot needed to arrest growth, but probably more than 250 g total berry weight per shoot will be needed to control growth usefully. Low-cropping varieties may have inadequate yield to limit vigour sufficiently.

We can explain this growth reduction by saying that growing berries act as *sinks* or attractants for the carbohydrates and other materials produced in leaves. If the demand from the berries is high, they are diverted from shoot tips, which cease growth. In addition, such materials are directed away from the roots so that all parts of the plant are devigorated.

By contrast, shoots with no berries are invigorated because of the excess carbohydrates available. Growers, especially in vigorous situations, should therefore reduce the number of shoots after 20–30 cm of growth in spring, paying particular attention to removal of unproductive shoots.

ROOT PRUNING AND RESTRICTION

As noted earlier, the plant tends to create a specific ratio between weight of the root and that of the above-ground parts. After hard pruning, the plant encourages the growth of strong shoots so as to re-establish the balance. Root pruning (if we were able to do this) would discourage shoot growth by returning the root/shoot ratio to normal. Some growers have run a ripper along the sides of rows to prune the roots, but the results generally have been too variable to encourage more growers to do this.

Roots that are restricted in some way may also result in reduced shoot growth. Some vineyards do have a naturally restricted root run to assist in growth control. Research workers are presently experimenting with permeable bags positioned underground. These are designed to restrict root growth but allow flow of water and nutrients. They can be successful, but the economics of the operation have still to be determined.

FIGURE 45: Sudan grass being used for vigour control in a Washington State vineyard. Note the windmill for frost control in the background.

COMPETITIVE CROPS

Other plants are sometimes grown between the rows to reduce vine vigour. These have varied from grass to clover and from chicory to Sudan grass (Figure 45) or just uncontrolled weed growth. The aim is to remove the availability of water and nutrients from the vine.

SHOOTS KILLED BY FROST IN SPRING

In the unfortunate situation where a spring frost has killed the new shoots and where there is little likelihood that an economic crop will come from the next flush of shoots, growers will be faced with the problem of managing the vines for the remainder of the season.

After frost damage, return shoot growth will tend to concentrate around the head of the vine. On the canes, many of last year's buds will not regrow and gaps will appear in the canopy. To ensure enough replacement canes (in cane-pruned vines) but to overcome excess vigour at the head, it is generally wise to remove about two thirds of the old cane after the frost. The remainder is removed in winter.

Persistent suckers and shoots

Normally, once shoots on the trunk have been rubbed off in the first two to three years after planting, further shoot production from this area is rare. If shoot growth persists, it is probably because of inadequate bud and shoot removal in these early years.

Buds at the base of shoots are very close together, and the lowest ones may not be obvious. If, in removing the shoots, we do not remove these buds, they retain the ability to grow in this or subsequent seasons, thus, every time we fail to remove a shoot adequately from the trunk, we increase the potential growing points, accentuating the problem.

The solution is to be thorough in shoot removal. This can be by close cutting at the base of the shoot or by rubbing off buds shortly after bud burst. Even breaking all shoots by hand is better than cutting and leaving a stub (see Figure 46, above).

Chemical removal is possible. Spraying some non-systemic herbicide, such as paraquat, on the growing suckers kills the tips and the young leaves. A single spray is insufficient because the lower buds will be retained. Failure to repeat sprayings exaggerates the problem. Spraying must continue till regrowth ceases to occur; take care, chemicals such as paraquat are very dangerous and need careful application.

FIGURE 46: *Above*: incorrect removal of suckers at base of trunk. This form of pruning will increase the numbers of suckers next year. *Below*: correct removal of shoot on trunk.

Trouble Shooting

Badly bowed trunks

The situation where trunks were straight but have badly bowed is especially common where a string is used to provide a straight trunk (see Figures 36, page 45; 48, page 63). While this method is satisfactory in the first year, the weight of the vine, as it begins to crop, pulls down the wire, especially towards the centre, and the trunks in that area bend. This scenario is much worse when intermediate posts are far apart. Once the trunk is fixed in this position, it is difficult or impossible to rectify.

The solutions are:
- to use a stake rather than a string,
- place a small prop with a notch or nail midway between the intermediate posts to prevent sagging (it can be removed later in the life of the vineyard),
- not to have the intermediate posts too far apart.

11 Establishing and Training Young Vines

Getting the vine to the stage where it will be pruned and trained as described above takes several years. The amount of time depends on several factors (described shortly). This chapter discusses the methods used for early pruning and training of young vines.

Planting material

Vines may be established using cuttings, rooted cuttings, or grafted plants with roots. Unrooted or rooted cuttings are recommended only where phylloxera or nematodes are not problems. Rooted cuttings or grafted plants are generally preferred because they establish quickly and sturdily and fewer gaps occur because of plants not surviving. Nevertheless, some growers have successfully established vineyards using unrooted cuttings.

Death of cuttings almost always occurs because insufficient water is taken up by the stem end and/or developing roots. It is therefore vitally important that the soil does not dry out. However, irrigation may cool the roots, slowing their growth, while shoots may grow vigorously because of warm, ambient temperatures. Watering therefore must not be excessive. It should be sufficient to maintain a moist environment but not to over-soak the root zone.

Some growers place cuttings in cool store until spring when the ground has had the opportunity to warm up. Alternatively, nursery staff can place cuttings in a hot bed to provide the grower with cuttings that have callused and where a few roots are just beginning to appear. Planting through black polythene reduces moisture loss and helps warm the soil beneath the plastic.

We need to emphasise to growers that planting material must be of top quality. This stipulation means: cuttings at least pencil thickness at the top, rooted plants well grown and of adequate vigour, high-health material not carrying viruses or other pathogens, and appropriate clones of the desired cultivar. Very importantly, the material must never be allowed to dry out before planting. Plants and cuttings quickly lose moisture if exposed to warm air. Such plants will establish less effectively or not at all.

Usually, plants are cut back to two or three obvious buds before planting. All other buds must be carefully removed.

Ground conditions

While it has not been our intention to cover soil management in this monograph, a few aspects related to planting and establishment of vines are pertinent:

- Growers should ensure the soil is weed-free before planting. It is especially important to get rid of any perennial weeds that can regenerate from underground roots or stolons.
- The soil in which the grapes are to be planted should be cultivated and firm. If left too loose, air may permeate to the root zone, encouraging drying of roots.
- Fertiliser or lime additions may be required to correct soil deficiencies. Materials like lime or phosphates are best incorporated in the soil before planting.

Wind, water and weeds

Competition from weeds is the most common reason why vineyards are slow to establish. Even if growers think they have provided adequate water and fertiliser, severe weed competition will still limit growth. Wind and lack of water cause dehydration. Wind may also break inadequately staked vines. Protection from rabbits and other vermin may also be required.

Planting

Cuttings may be placed with only their top third above ground. Removal of all but the top two buds reduces suckers later. Rooted cuttings and grafted vines are planted at the previous ground level. Most importantly, the graft union should be at least 10 cm above the ground surface.

Growers can adopt many methods of planting. Hand planting offers the type of care given to a choice garden shrub, although time

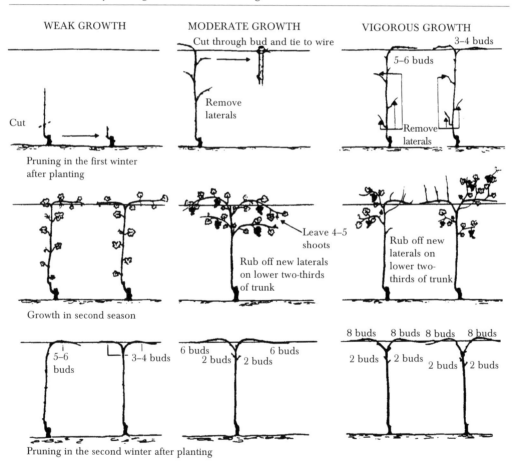

FIGURE 47: Early training of vines of different vigour.

usually dictates quicker methods. Cuttings may simply be pushed into the ground or a pointed stick used to make the initial hole. Water jets can also be used to make the hole for cuttings or rooted plants. Large vineyards sometimes use planting machines. After planting, the soil around the plant must be firmed (but not compacted). If the soil is dry, moisture that is sufficient to wet but not soak the root zone should be applied.

Training young vines

Despite the fact that established vineyards may not be well served by vigorous growth, some vigour at establishment is wise. Growers therefore may not wish to fertilise the ground generally but to add some to the planting hole or the irrigation water (fertigation). While some moisture stress may be a management tool for the established vineyard, it is not good practice for the young vine. Speed of establishment makes economic sense to reduce the time of negative cash flow.

One exception centres on the situation where, under very vigorous conditions, internodes may be too widely spaced to provide the appropriate number of buds near the head of the vine. Here, vigour control may be needed to reduce internode length.

Figure 47 shows a training programme for the first two seasons after planting under weak, moderate and vigorous conditions. The plant has been planted with two to three buds retained. The best shoot from these buds has been selected and is trained upward towards

FIGURE 48: Supporting the shoot.

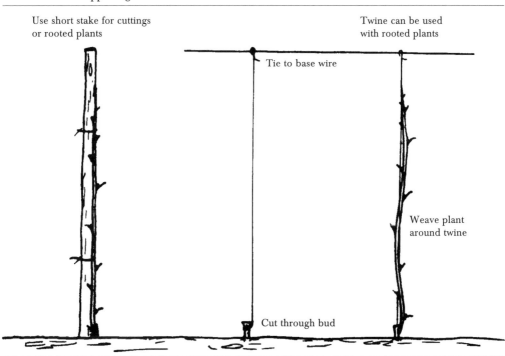

Establishing and Training Young Vines 63

the first wire. It is supported by a stake or string as shown in Figure 48.

Sometimes in the first year, the shoot will not reach the wire, in which case it is normally pruned back the next winter to three or four buds. If growth is very vigorous and reaches the wire early in the season, it may either be trained one way along the wire or cut off below the wire to encourage the shoot to divide. If growth is moderately vigorous and reaches the wire later in the season, it will be cut off in winter and tied to the wire.

Subsequent treatments of weak, moderate or vigorous vines are shown in the lower part of Figure 47. Here, vines with moderate to vigorous growth have produced a small crop in the second season after planting. After the second winter – the third growing season – these fruiting vines are nearing full production. Weak vines will need a further season to reach this stage.

The trellis

We have not detailed the technical aspects of trellis construction in this monograph. Most districts have contractors or advisers who can provide on-the-spot assistance. For those wishing to obtain a technical overview of this topic, we recommend Smart and Robinson's book *Sunlight into Wine* (1991). Trellis construction is vitally important, which means that growers must carefully check advice given and not skimp to reduce costs. Failure to heed this advice risks an expensive gamble.

References

Archer, E and Strauss, H C (1989) 'Effect of shading on the performance of *Vitis vinifera* L cv Cabernet Sauvignon'. *S Afr J Enol Vitic*, 10: 74–77.

Carbonneau, A (1979) 'Research on criteria and outlines of training systems for the grapevine: extension to woody perennial plants'. *Ann Amelior Plantes*, 29: 173–185.

Champagnol, F (1984) *Elements de Physiologie de la vigne et de viticulture generale*. Publ the author, Montpellier.

Gladstones, J (1992) *Viticulture and Environment*. Winetitles (Adelaide).

Jackson, D I (1983) 'The Lincoln canopy for grapes'. *Am J Enol Vitic*, 34: 176–79.

Jackson, D I (1988) 'Prediction of a district's grape-ripening capacity using a latitude temperature index (LTI)'. *Am J Enol Vitic*, 39: 19–28.

Jackson, D I and Cherry, N J (1988), 'Prediction of a district's grape-ripening capacity using a latitude temperature index (LTI)'. *Am J Enol Vitic*, 39: 19–36.

Jackson, D I and Lombard, P B (1993) 'Environmental and management practices affecting grape composition and wine quality: a review'. *Am J Enol Vitic*, 44: 409–30.

Jackson, D and Schuster, D (1994) *The Production of Grapes and Wine in Cool Climates* (3rd ed). Christchurch: Gypsum Press.

Lavee, S and Haskal, A (1980) 'An integrated high density intensification system for table grapes'. Int Symp, Centennial Univ Calif at Davis, 390–98.

Morrison, J C and Noble, A C (1990) 'The effect of leaf and cluster shading on the composition of Cabernet Sauvignon grapes and on fruit and wine sensory properties'. *Am J Enol Vitic*, 41: 193–200.

Shaulis, N J, Amberg, H and Crowe D, (1966) 'Response of Concord grapes to light exposure and Geneva Double Curtain training'. *Proc Am Soc Hort Science*, 89: 268–80.

Smart, R E (1973) 'Sunlight interception in vineyards'. *Am J Enol Vitic*, 24: 141–147.

Smart, R E (1987) 'Influence of light on composition and quality of grapes'. *Acta Hort*, 206: 37–47.

Smart, R E and Robinson, M (1991) *Sunlight into Wine*. Adelaide: Winetitles.

Smart, R E Smith, S M and Winchester, R V (1988) 'Light quality and quantity effects on fruit ripening for Cabernet Sauvignon'. *Am J Enol Vitic*, 39: 250–58.

Van den Ende, B and Chalmers, D (1983) 'The Tatura Trellis as a high density system for pear trees'. *Hort Science*, 18: 946–47.

Index

Abscission   5, 8
Acids   5, 15
Alpha zones   2–3
Apical dominance   8
Aroma of wines   15

Beta zones   2–3
Bud numbers   32, 33
Bud number, estimating   34
Bud number recommendations   38–39, 43–47, 49, 51, 54
Bud position on cane   13
Buds per hectare   32
Buds per metre of row   32
Buds, 'count'   25
Buds, blind   25
Buds, compound, primary, secondary, tertiary   6–7
Buds, dormant   12

Cane pruning   25–29
Canes, arched   29
Canopy height   19
Capfall   14, 40, 43, 47, 57
Carbohydrates   4
Carbon dioxide   4, 6
Cellulose   4
Chlormequat (CCC, 'Cycocel')   57
Chloroplasts   4
Classical vineyards   33
Climates, cool   3, 16
Climates, warm   3, 17
Colour development   15
Competitive crops for vigour control   58
Composition of juice and wine   14
Configurations for vine trellises   36
Cool climate viticulture   2
Cool climates, definition   3
Cordon cane   28, 29
Coulure   13

Cropping potential   16–17
Cropping to control vine vigour   58
Cultar   57
Curtain training   42–43

Decision making   54
Degree days   3
Dependent and fixed variables   40
Distance apart of rows   19
Double head   29
Double vertical plane   36

Early Bunch-Stem Necrosis (EBSN)   14, 18
Early Growth Principle (EGP)   10
EBSN   14, 32
End Point Principle (EPP)   8, 25
Establishing and training young vines   62–64
Ethephon   58
Ethrel   57

Fertigation   63
Fertilisation   14
Flavour components   15
Flowers   12
Frost, consequences   58–59
Fructose   4, 6
Fruit development   14
Fruit set (coulure)   13
Fruitfulness   13

Geneva Double Curtain (GDC)   48–49
Girdling of trunks   56–57
Glucose   4, 6
Glycogen   4
Goblet system   18, 35
Gravity   9
Guard cells   5

Head cane with canes both sides   27–28
Head cane with canes one side   28–29
High Sylvoz   42–43

High-density planting   30
Highest Point Principle (HPP)   9, 28

Inflorescence   6, 7, 12, 13
Initiation of inflorescences and flowers   3, 13, 32
Internodes   6, 9
Irrigation   63

Latitude   21–22
Latitude Temperature Index (LTI)   3
Leaf removal   33
Leaf-fruit ratio   14
Light   4, 14
Light intensity   5, 13
Light interception and utilisation   6, 18, 20, 21
Light penetration   3
Light quality and quantity   5
Lincoln canopy   22–23
Low-density planting   30–31
Lyre   51–52

Macroclimate   2, 15
Magic numbers   33
Maturation of fruit   14
Maturity, lateness of   3, 15
Mesoclimate   2, 15
Metabolism   6
Methoxypyrazines   15
Microclimate   2, 15
Mid-height Sylvoz   43, 44–47
Mid-level cordon with downward–pointing canes   44–45
Mineral uptake   5
Minerals   4
Moselle system   18, 53
Must   5

New world vineyards   33
Nodes   6, 7
Nodes per hectare   32
Non-vertical canopy   36

Oxygen   6

Pergola   22–23
Pests and diseases   18
pH   5, 15
Phenolics   15
Photosynthesis   4, 5, 14, 15, 33
Planting   63–64
Planting material   61
Planting, high density   30–31
Planting, low density   30–31
Pole, training on   53
Pollination   14
Positioning   3
Potassium   5, 15
PP333   57
Principle of the node-trunk ratio (PONT)   10
Pruning/training strategies   24

Quality of wine   14

Rainfall   14
Respiration   6
Root pruning and restriction   58
Root system   12
Root-Shoot Principle (RSP)   10
Row direction (orientation)   20, 22
Row width   19, 30
Ruakura twin two tier   52

Scott-Henry   46–47
Senescence   5
Set   3
Shade   3, 14, 33
Shoot growth   11, 12
Shoot variation   55, 56
Shoots   59
Shoots killed by frost   58
Single upright with downward-pointing shoots   42–47
Single vertical plane   36
Soil management   62
Spur pruning   25–29
Starch   4
Stomates (or stomata)   5
Strategies for survival   11–13

Suckers   59
Sucrose   4
Sugar levels   14
Sugars   6

Tatura trellis   22–23
Temperature   13
Temperatures, cool   14
Temperatures, hot   21
Tendrils   7, 8, 11
Topping   3, 40
Training systems   3
Training young vines   63–64
Translocation   6
Transpiration   5–6
Trellis   3, 64
Trellises in the double vertical plane   48–54
Trellises in the single vertical plane   37–47
Trellising   35–54, 64
Trimming   3, 33
Trouble shooting   55–60
Trunk Proximity Principle (TPP)   9, 28

Trunks become bowed   60
Tying cane to wire   29

Variables – dependent and fixed   40
Varieties – cropping potential   16–17
Vegetative growth   6
Vertically Shoot-Positioned canopy or VSP   18, 37–41
Vigour – systems suited to   52
Vine spacing   30–31
Vines are too vigorous   56
Vineyard configuration or geometry   18
VSP   18, 37–41

Water   4, 5, 6, 63
Weeds   62
Wind   62
Wire Care-Clip   38
Wires, high tensile   36, 37

Xylem   5

Yield potential   16, 35–54